*Implications of*
*Anti-Ballistic Missile Systems*

PUGWASH 

# Implications of Anti-Ballistic Missile Systems



PUGWASH MONOGRAPH II

Edited by
C. F. BARNABY & A. BOSERUP

THE HUMANITIES PRESS · NEW YORK

First published in the United States of America 1969
by Humanities Press Inc,
303 Park Avenue South, New York, N-Y 10010



Printed in Great Britain

# Contents

## Part I

## INTRODUCTION

### *Anti-Ballistic Missiles and their Deployment*



## Part II

## PAPERS PRESENTED TO THE SYMPOSIUM

### *Section A: Implications for the Arms Race*

## *Section B: Political Implications*



## *Section C: Strategic Implications*



## *Section D: General Implications*

## Part III

## SUMMARY OF DISCUSSIONS

# List of Participants

| | |
|---|---|
| H. AFHELDT (F.G.R.) | Law, International Relations: Secretary of the German Federation of Scientists. |
| C. F. BARNABY (U.K.) | Physics: Executive Secretary of Pugwash Conferences on Science and World Affairs. |
| A. BOSERUP (Denmark) | Physics: Institute for Peace and Conflict Research. |
| F. CALOGERO (Italy) | Physics: Professor, Rome University. |
| D. CARLTON (U.K.) | International History: London. |
| B. T. FELD (U.S.A.) | Physics: Professor, Massachusetts Institute of Technology. |
| S. FREIER (Israel) | Physics: Deputy Director of Weizmann Institute of Science. |
| A. J. R. GROOM (U.K.) | International Relations: London University. |
| V. HAJDU (Czechoslovakia) | International Law: Academy of Sciences, Prague |
| N. HERLOFSON (Sweden) | Physics: Professor, Royal Institute of Technology. |
| J. KLEIN (France) | Social Science: Centre d'Etudes de Politique Etrangère. |
| M. LEITENBERG (SIPRI) | Biochemistry: International Institute for Peace and Conflict Research. |
| O. MAALØE (Denmark) | Molecular Biologist: Professor, University of Copenhagen. |
| G. L. MAGNUSSON (Sweden) | Engineering: Research Institute of National Defence. |
| I. V. MILOVIDOV (U.S.S.R.) | Engineering: U.S.S.R. Academy of Sciences. |
| K. MORTENSEN (Denmark) | Law: Danish Ministry of Foreign Affairs. |
| R. NEILD (SIPRI) | Economics: Director of International Institute for Peace and Conflict Research. |
| M. NITA (Rumania) | Professor, Aeronautics Commission, Academy of Sciences. |

| | |
|---|---|
| N. PETERSEN (Denmark) | Social Science: Danish Institute of International Affairs. |
| I. G. POCHITALIN (U.S.S.R.) | U.S.S.R. Academy of Sciences. |
| J. PRAWITZ (SIPRI) | Nuclear Chemistry: International Institute for Peace and Conflict Research. |
| G. W. RATHJENS (U.S.A.) | Political Science: Professor, Massachusetts Institute of Technology. |
| P. B. REHBERG (Denmark) | Zoophysiology: Professor, Chairman of Atomic Energy Commission. |
| J. ROTBLAT (U.K.) | Physics: Professor, London University. (Secretary-General of Pugwash Conferences on Science and World Affairs). |
| J. J. STONE (U.S.A.) | Mathematics: Stanford University. |
| A. P. VINOGRADOV (U.S.S.R.) | Geochemistry: Academician, U.S.S.R. Academy of Sciences. |
| N. WIERDA (Netherlands) | Member of Parliament, The Hague. |

The following were unable to participate in the Symposium but submitted papers which have been included in this book:

| | |
|---|---|
| J. DELBRÜCK (F.G.R.) | International law: Institute of International Law, Kiel University. |
| D. R. INGLIS (U.S.A.) | Physicist: Argonne National Laboratory |
| F. A. LONG (U.S.A.) | Chemist: Professor, Cornell University. |

# Part I

# INTRODUCTION

## *Anti-Ballistic Missiles and their Deployment*

C. F. BARNABY

# 1: The Development and Characteristics of Anti-Ballistic Missile Systems

There is no doubt that the implications of the present deployment of Anti-Ballistic Missile (ABM) systems and of the future decisions which must inevitably be taken regarding the ultimate levels of these systems will have major consequences for world security. The accurate evaluation of ABM systems is beset by many doubts and uncertainties. Although this makes it difficult to predict with confidence the effectiveness of ABMs it also means that discussions and analysis of the possible effects of ABM deployment are of particular value. In such a situation it is inevitable that there should be fierce debate about the various issues raised by the development of ABM systems. As in most aspects of International Relations, there is no absolute truth in these issues but this does not mean that every suggestion is equally reasonable, in fact, it is in just this situation that careful and considered analysis proves its worth. Such analysis is even more urgent in the case of the implications of ABM deployment because, in the ultimate analysis, decisions will probably be made (and have been made) for largely psychological and political reasons rather than for strategic ones. International Politics is, in any case, affected more by perceptions than facts, and in a situation when many of the underlying issues are of a technological nature, difficult for the non-scientist to understand fully, and when there are many uncertainties associated with an accurate evaluation, perceptions are likely to become even more predominant.

It is noteworthy that the Soviet ABM deployment caused relatively little comment, possibly because the Soviet emphasis on defence made such deployment a not very surprising development. Now that both superpowers are engaged in ABM deployment, and that the possibility of ABM systems is being considered for other areas, the ramifications involved are receiving greater attention. Even so the debate on the issues now taking place in the West is largely confined to the United States and it is important that those in third countries are stimulated to take part in the debate and consider their attitudes to these issues in a relaxed atmosphere. It is only in this way that hasty, and therefore probably incorrect, policies will be avoided by the decision-makers. In particular, it is necessary to consider the political consequences of the possible installation of ABM systems in other parts of the world—for example, in Western Europe.

Finally, discussion of the consequences of the development and deployment of new weapons such as the Multiple Independently-Targetable Re-entry Vehicles (MIRVs) and the Fractional Orbital Ballistic System (FOBS) is also urgently required.

## DEVELOPMENT OF ANTI-BALLISTIC MISSILE SYSTEMS

Anti-Ballistic Missile systems are designed to detect, identify, intercept and destroy the warheads of hostile missiles.

### *United States ABM Development*

American research to produce a defence against ballistic missiles was begun by the United States Army in the mid-1950s. The Nike-Hercules anti-aircraft system (deployed in 1958) was developed into the Nike-Zeus anti-missile system. In 1959 the Army recommended the procurement of the Nike-Zeus System, to be operational in 1963 or 1964. The proposal was rejected but research and development continued. Meanwhile the U.S. Air Force, whose rival ABM system called Wizard was cancelled, was given the task of developing a missile warning radar network. As a result the Ballistic Missile Early Warning System (BMEWS) was developed and installed during the period 1960–3. The Missile Detection and Alarm

System (MIDAS), which uses satellites to detect missile launching by infra-red radiation, is being developed to supplement the BMEWS stations. At present there are BMEWS stations in Alaska, Greenland and England. Apart from BMEWS, the United States has two chains of radar and tracking stations known as the Pinetree Line and the Distant Early Warning Line (DEW). The Semi-Automatic Ground Environment System (SAGE), organized at 13 stations within the United States, co-ordinates the tracking of objects in the air space and 11 of these stations are to be linked with Back-Up Interceptor Control (BUIC) stations. Some coastal radar stations are designed to detect submarine-launched missiles.

Early in 1963 it was decided not to procure the Nike-Zeus system, which had a limited range and kill radius, because its effectiveness against a large-scale attack was in doubt even though the Army had conducted a series of 14 single intercept-tests with 70 per cent success. Instead, it was decided to proceed with the development of a more advanced system, the Nike-X. This system included major advances in radar technology which increased the amount of traffic the system could handle. The earlier mechanical radars were replaced by phased-array radars supported by large modular computers able to track many targets simultaneously. The Zeus-missile was also supplemented with a new high-acceleration interceptor missile called Sprint. Sprint, designed to intercept enemy Inter-Continental Ballistic Missiles (ICBMs) which escape the Zeus missiles, has a high acceleration of over 100 g and a short range of between 30 and 50 km.* It is an 8-metre two-stage solid-fuelled missile which is ejected from its cell by a piston activated by a gas generator. After ejection the first stage ignites sending the missile to an altitude of 20 km within about 4 seconds. Sprint has a nuclear warhead with a yield equivalent to several tens of kilotons of TNT. The Nike-X system, however, still remained a 'point defence', i.e. able to

* The altitude of Sprint interception may be as low as 10–20 km; this would be well within the atmosphere and only a few seconds from ground impact. If the incoming warhead was timed to explode at a high altitude the Sprint might not be able to intercept it before the explosion.

protect only a single target by engaging incoming ICBMs in a 'bullet-to-bullet' duel. To provide an 'area defence' which would protect large areas (measured in thousands of square kilometres), including the cities and strategic forces within them, requires a long-range missile with a large warhead that is able to engage incoming ICBMs above the atmosphere. In 1965 the United States began the development of a long-range missile, called Spartan, to replace the Zeus missiles in the Nike-X system. The Spartan missile can intercept an incoming missile at an altitude of up to 450 km and at ranges of up to 650 km from its location. The warhead of the Spartan missile, normally greater than one megaton, mainly relies upon the so-called 'Zapp effect' which uses the surge of soft X- and other radiation produced by a high-yield nuclear explosion. The majority of the total energy produced by the explosion escapes in the form of such radiation; this can penetrate thousands of miles in the near vacuum of space.

### *Methods of ICBM Destruction*

Both the Spartan and Sprint missiles would destroy an incoming ICBM warhead by means of a nuclear explosion. The warhead would, however, usually not be destroyed by blast, but, instead, would be inactivated in one of several different ways depending on: the height of the ABM detonation; its distance from the incoming missile; and the design and size of the ABM warhead. Firstly, thermal radiation may either damage the heat shield of the hostile warhead so that it burns up on re-entry or damage the electrical systems within the warhead. Secondly, the surface of the warhead may be subjected to a high electric charge and the accompanying electromagnetic shock-wave may damage the detonating mechanism in the warhead. Thirdly, high-energy neutrons produced by the nuclear explosion of the ABM may either prematurely fuse the nuclear core of the warhead or effectively reduce the critical mass in the warhead. Fourthly, exposure to high-energy gamma-radiation may damage the electronic circuits of the guidance system and/or the detonating mechanisms in the warhead. In fact, electronic components can be damaged by exposure to X-rays, neutrons and other radiation. The hostile warhead could thus be rendered useless by physical damage to the heat shield, or by premature fusing, or by damage to the electronics of the guidance and detonating systems. An ABM system employing long-range missiles using the Zapp effect can provide an area defence covering thousands of square kilometres with relatively few anti-missile batteries.

An ABM system used for point defence (i.e. of areas having radii of a few tens of kilometres) would employ defensive missiles designed to destroy or incapacitate the incoming warhead in the terminal stage of its trajectory after it has re-entered the atmosphere. The interception could occur at distances of less than about 30 km from the target of the incoming missile. The missiles used for point (or terminal) defence, such as Sprint missiles, need have only a relatively short range but they do need very high acceleration. If the detonation of the warhead occurred within a short distance (a few hundred metres) of the incoming missile it is possible that the blast from the explosion could destroy the enemy missile since the interception would occur in the atmosphere: close interceptions could also result in the destruction of the opponent's missile by the neutron effect. The latter mechanism would be effective over a greater distance than the blast effect but, because relatively intense neutron fluxes would be necessary to incapacitate the warhead of the incoming missile, it is unlikely to be successful for interceptions in the atmosphere at distances greater than a few tens of kilometres. Therefore, interceptions at long ranges (of hundreds of kilometres) must rely on the effects of X- and gamma-radiation which have great power of penetration within the atmosphere.

*Radar Systems*

The highly complex and expensive radar system associated with the United States ABM system includes: the Perimeter Acquisition Radar (PAR), a low-frequency radar which detects the missile at a long range, tracks the missile and predicts its path; the Missile Site Radar (MSR), a higher frequency radar, which guides the Sprint and Spartan missiles from each battery; the Multi-function Army Radar (MAR) which is a powerful phased-array radar designed to perform all the functions associated with countering a full-scale attack – i.e. detect the hostile targets, select the warheads from decoys, track the trajectories of the warheads and direct the interceptors; and the TACMAR radar which is a smaller version of MAR having a lower target handling capacity. The MAR system is controlled by a complex computer system and is composed of many individual separate units which can be combined into systems of different sizes. All of these radars are based on phased-array technology whereby many beams are generated and instantaneously moved across the sky. Phased array radars have a scanning angle limited to about 90° of azimuth whereas mechanical radars, which they replace, can scan through a full 360° of azimuth. This disadvantage can however, be overcome by deploying more than one transmitting and receiving array at each site.

*Cost of ABM System*

The approximate costs of an ABM system of the type described above have been given by Mr McNamara (the American ex-Secretary of Defense) for two alternative deployments, a $9 million programme (Posture A) and a $20 million programme (Posture B). Posture A represents

> a light U.S. defence against a Soviet missile attack on our cities. It consists of an area defence of the entire continental United States, providing redundant (overlapping) coverage of key target areas; and, in addition, a relatively low-density Sprint defence of a number of the largest cities to provide some protection against those warheads which get through the area defence.

Posture B is

> a heavier defence against a Soviet attack. With the same coverage it provides a higher Sprint defence for twice the number of cities.

Posture A would provide a Sprint defence for the 26 largest cities and Posture B for the 50 largest cities. The costs given by

McNamara, who warned that the estimates may be understated by 50 to 100 per cent, are given in Table 1. If the quoted Atomic Energy Commission (AEC) cost is for the provision of warheads the total number of interceptor missiles envisaged is about 2000 and 4000 for Posture A and B respectively since the cost of Sprint and Spartan missiles is likely to be of the order of $1 million each.

TABLE I

*U.S. ABM Deployment Costs*

| | *Posture A Investment Cost Billion $* | *Posture B Investment Cost Billion $* |
|---|---|---|
| Radars (MAR, TACMAR, PAR, MSR) | 6·5 | 12·6 |
| *Missiles* | | |
| Spartan and Sprint | 2·4 | 4·8 |
| DOD (Department of Defence) | 8·9 | 17·4 |
| AEC | 1·0 | 2·0 |
| Annual Operating Cost | 0·38 | 0·72 |

McNamara pointed out that the populations of the unprotected, or relatively unprotected, areas of the United States would apply political pressure for equal protection with, say, New York and Washington, and for this and other reasons he believed that an ABM system deployed with the objective of protecting the United States against an attack from the Soviet Union would require an expenditure of at least $40 billion over a ten-year period. The American Joint Chiefs of Staff have, however, recommended the procurement of Posture A and they are strongly supported in Congress and by the defence industry. It is estimated that about five years would be required to install the system.

### *Cost-exchange Ratios*

The cost-exchange ratio between defensive and offensive force increases is sometimes used to judge the potential arms race implications of ABM systems. This ratio is the cost of an ABM system divided by the cost of the increased enemy offensive forces which would offset the ABM system. Figures for the cost-exchange ratio have been given by McNamara and these are shown in Table 2.

TABLE 2

*Cost-exchange Ratios*

| *Level of U.S. prompt fatalities (i.e. by blast and fall-out) which the Soviets believe will deter a U.S. first strike* | *Cost-exchange ratio (i.e. American ABM cost divided by Soviet offsetting offensive forces cost)* |
|---|---|
| *millions* | |
| 40 | 4 : 1 |
| 60 | 2 : 1 |
| 90 | 1 : 1 |
| 100 | Undefended case |

The calculations assume a certain ABM effectiveness and presume a substantial United States ABM system and a Soviet second strike; the figures must, therefore, be used with caution. It will be seen, however, that the lower the level of perceived deterrence the more favourable is the ratio between the cost of the offsetting forces and the cost of the defensive forces. Mr McNamara has said that the United States' assured destruction capabilities must guarantee the destruction of between 20 and 25 per cent of the Soviet population and between 50 and 70 per cent of the industrial capacity of the Soviet Union; he also has credited the Soviet Union with the potential capacity to destroy more than 50 per cent of the United States population. If, however, the Soviet Union, for example, perceived that assured destruction capacity of, say, 25 per cent the United States population was a sufficient deterrent, then the required expenditure on offensive forces to offset an American ABM system would be about one quarter of the American ABM cost. In other words, it is not necessary that one superpower should react to the ABM system of the other superpower by increasing its offensive forces to the point where the cost-exchange ratio was equal to or less than one. It should be emphasized that many arguments based on these figures (which are themselves open to criticism) ignore vital points such as the tendency to overestimate the effectiveness of an opponent's ABM system and so on.

The quantitative element introduced into ABM discussions by the use of cost-exchange ratios are usually regarded as illusory. Experience has shown that initial estimates of the cost of any new weapon system are usually grossly underestimated. Moreover, it is not feasible to test the performance of an ABM system under realistic operational conditions. By its nature an ABM system would have to operate in an environment produced by many nuclear explosions and the effect of these on the various components of the system cannot be pre-judged or tested on the scale at which they would occur. Furthermore, the question of the level of damage and casualties in superpower A which would be regarded by superpower B as sufficient to deter superpower A involves a value judgment involving situations which the human mind is hardly capable of envisaging, for example the effect of several tens of millions of casualties and the consequences of reducing all, or most, of a society to a radioactive desert. Consequently, it is not sensible to regard cost-exchange ratios as realistic quantitative indicators on which to base arguments.

In spite of these considerations, McNamara has given estimates (see Table 3) of the prompt casualties in a total strategic exchange. The estimates assume that ABM systems would perform at a high level of effectiveness. If it is also assumed that there is no Soviet reaction to an American ABM deployment the difference in American casualties between an American first and second strike is 10 million; this is 10 million less than the undefended differences. Similarly, the estimates given for American casualties assuming Soviet reaction to American ABM deployment are used to argue that ABM deployment is not conducive to pre-emptive stability. It must be emphasized, however, that the assumptions made to arrive at the estimates and the great uncertainties in the performance of ABMs are

TABLE 3

*McNamara's Estimate of the Number (in millions) of Fatalities in a Total Strategic Exchange*

(1) *Assuming no Soviet Reaction to the U.S. ABM Deployment*

| Programme | Soviets strike first U.S. retaliates | | U.S. strike first Soviets retaliate | |
|---|---|---|---|---|
| | *U.S. fatalities* | *Soviet fatalities* | *U.S. fatalities* | *Soviet fatalities* |
| Approved | 120 | 120 + | 100 | 70 |
| Posture A | 40 | 120 + | 30 | 70 |
| Posture B | 30 | 120 + | 20 | 70 |

(2) *Assuming Soviet Reaction to the U.S. ABM Deployment*

| Programme | Soviets strike first U.S. retaliates | | U.S. strike first Soviets retaliate | |
|---|---|---|---|---|
| | *U.S. fatalities* | *Soviet fatalities* | *U.S. fatalities* | *Soviet fatalities* |
| Approved | 120 | 120 + | 100 | 70 |
| Posture A | 120 | 120 + | 90 | 70 |
| Posture B | 120 | 120 + | 90 | 70 |

such that calculations based on the estimates should be treated with caution.

*United States Sentinel System*

In a speech in September 1967 McNamara announced the United States Government's decision to deploy a 'thin' ABM system (called Sentinel) designed primarily to protect the United States against a hypothetical Chinese ICBM attack in the mid-1970s. McNamara stated that although both the United States and the Soviet Union possessed 'second-strike capability' against each other, neither possessed 'first-strike capability'. He defined 'second-strike capability' as the capability to absorb a surprise nuclear attack and survive with sufficient power to inflict unacceptable damage on the aggressor. 'First-strike capability' was defined as 'the substantial elimination of the attacked nation's retaliating second-strike forces'.

In McNamara's opinion the fact that the United States had a substantial nuclear superiority over the Soviet Union 'of at least three or four to one' was in itself of limited significance. 'The blunt fact is that neither the Soviet Union or the United States can attack the other without being destroyed in retaliation, nor can either of us attain a first-strike capability in the foreseeable future.' The United States, however, has 'an overwhelming first-strike capability with respect to China and the power not only to destroy China's nuclear power completely but to devastate her society as well'. McNamara also said, 'indications are that China will have medium-range ballistic missiles within a year or so, an initial intercontinental missile capability in the early 1970s, and a modest force in the mid-1970s'. The deployment of 'a relatively light Chinese-oriented ABM system' did not mean that this would be expanded into 'a heavy Soviet-oriented ABM system' which would 'in effect be no adequate shield at all against a Soviet attack but rather a strong inducement for the Soviets to vastly increase their own offensive forces'. He described the Soviet ABM system as 'light and modest'. The thin system was said to be 'relatively inexpensive' – preliminary estimates place the cost as about $5 billion and an additional $500 million a year to operate – and would have a much higher degree of reliability against a Chinese attack than the much more massive and complicated system that some have recommended against a possible Soviet attack. The first interceptor battery will be in operation in the early 1970s and the entire American system should be ready by 1973. Sentinel will consist of a relatively small number (15 to 20) of batteries of long-range Spartan missiles around the country. The aim of Sentinel is to provide protection for the entire country against an unsophisticated attack. The locations of the missile sites have not yet been finally decided but the State Department has chosen an initial ten areas for survey as possible sites. Whether or not Sprints are to be used is, at present, not clear. If they are used it appears that this will only be for the defence of key elements of the ABM system itself, in other words, for the point defence of radar and Spartan sites and not for cities or offensive missile sites.

It is planned that Sentinel should have six PAR installations along the northern boundary of the United States. An incoming missile on a minimum-energy ballistic trajectory would be sighted by PAR at a distance of about 4000 km, providing a warning time of about 10 minutes. After the incoming missile is sighted by PAR its trajectory would be tracked by MSR and this information allows the ABMs to be guided to the incoming missile using the computer system, the 'brain' of Sentinel, which has a data-handling and decision-making capacity greater than any previously constructed system. The MAR or TACMAR systems are not designed for use with a light ABM system such as Sentinel but would be installed only if the system were extended to counter a full-scale attack. In this connection it should be noted that Sentinel is designed to be of a flexible building-block nature, capable of extension into a heavier system into which the more complex radars and computers could be incorporated.

McNamara went on to claim that the Chinese-oriented ABM deployment would have a number of other advantages. It would provide 'an additional indication to Asians that we intend to deter China from nuclear blackmail, and this would contribute toward our goal of discouraging nuclear weapon proliferation among the present non-nuclear countries'. Further, the Chinese-oriented ABM deployment would enable the United States to add as a concurrent benefit a further defence of her Minuteman ICBM sites against Soviet attack which means that at modest cost it 'would in fact be adding even greater effectiveness to our [United States] offensive missile force and avoiding a much more costly expansion of that force'. As mentioned above, however, Minuteman sites are not at present to be protected by point defences. Finally, 'such a reasonably reliable ABM system would add protection of our population against the improbable but possible accidental launch of an intercontinental missile by any of the nuclear powers'. The official estimate about the effectiveness of the Sentinel deployment against any Chinese attack in the

1970s is that it would hold the consequential U.S. casualty figures 'at a level below one million'.

*Soviet ABM System*

ABM deployment by the Soviet Union probably began some years ago but it was not until the end of 1966 that Mr McNamara, then the American Secretary of Defense, publicly announced that the Russians were deploying an ABM system around Moscow, although United States press reports of ABM deployment by the Russians began in 1964. An ABM system may also be deployed around Leningrad. In addition, the Russians are installing a system along the north-western borders of the Soviet Union. The exact character of these installations, known as the Tallinn-Line, is uncertain, although in January 1968 the American Secretary of Defense said in Senate testimony that the Tallinn-Line did not have any significant ABM capability. It may be limited to anti-aircraft missiles. There have also been unconfirmed reports of activity around existing anti-aircraft establishments in the Soviet Union. The Russian system is presumably based on the short-range (40–50 km) Griffon missile and the long-range solid-fuel Galosh missile.

*Civil Defence Programmes*

Associated with ABM systems is a civil defence programme, including shelter provision, to which both the United States and the Soviet Union pay great attention. The Americans plan to spend 77·3 million dollars on their programme in the year 1969; the amount spent by the Soviet Union is not known with accuracy.

A shelter programme is necessary because, in a large-scale attack, some missiles would have a substantial likelihood of penetrating the ABM defences. For example, in the American system, warheads aimed at a target not defended by Sprint missiles may penetrate the outer Spartan defence. Also some warheads may detonate before the Sprint missiles could reach them. Such airbursts might yield significant blast and fire effects at ground level. As mentioned above, planned heights of interception may be less than about 15 km above the earth;

a five-megaton weapon exploded at this height would produce sufficient heat to ignite clothing at ground level and would probably start large-scale fires. In fact, ignition of clothing could take place in an area of radius of about 13 km from ground zero and second-degree burns in an area of radius of about 25 km from ground zero. Soviet ICBMs have warheads with yields of up to 30 megatons (see Table 6). Since the penetration, by incoming missiles, of both the area defence and the point defence is possible, fall-out shelters and blast shelters are essential for the protection of population centres. Estimates for the cost of a blast shelter programme for the United States vary from $38 billion for a system for 17 cities to over a $100 billion dollars for 50 cities.

*Nuclear Weapon Testing*

An important aspect of ABM development is the need for a new and comprehensive phase of underground nuclear testing with the purpose of developing ABM warheads. The United States announced twelve underground weapon tests in 1963, twenty-one in 1964, twenty-five in 1965, thirty-four in 1966 and forty in 1967. Thirteen Soviet underground tests have been announced by the USAEC and there may have been other unannounced tests by both powers. Pressure to resume atmospheric testing will no doubt increase, since information is required on the effect of nuclear explosions in producing interference with the operation of radars and computers. It is likely that the performance of the radar, communications and data-processing equipment associated with ABM systems will have to be tested against the effect of high-altitude nuclear explosions before any confidence can be put in the system.

*Response to ABM Deployment*

The deployment of an ABM system by one superpower would stimulate the other superpower to take steps to maintain his previously existing offensive capability. This could be achieved either by technological devices, called penetration aids, designed to improve the capability of the offensive missiles to deliver their warheads on to the target in spite of the ABM

TABLE 4

*Some possible Penetration Aids and Strategies*

| *Object* | *Method* |
|---|---|
| To confuse and disrupt radar linked to ABM system | Use of chaff, either light chaff (wire, foil) ejected during mid-course of trajectory or heavy chaff (chunks of metal) for re-entry phase;<br>Use of decoys;<br>By jamming;<br>By exploding a precursor nuclear warhead over territory of opponent;<br>By using slender pointed cone on warhead to reduce its 'visibility' on radar;<br>By use of anti-radar coating on missile |
| To evade ABMs | Harden warhead to withstand nuclear blast and radiation;<br>Use MIRVs;<br>Accelerate warhead down to target to cut the re-entry-to-target time;<br>Use manœuvrable re-entry vehicle |
| Other strategies | Explode high megatonnage warheads above range of ABMs used for point defence;<br>Saturate ABM system by attacking local areas with large number of missiles;<br>Explode high yield nuclear warheads at ground level down-wind of target where there are no ABMs and rely on fall-out to produce casualties |

system, or by developing attack strategies designed to by-pass or overcome the ABM system (Table 4).

Among the most practicable penetration aids are devices to harden the warhead of the strategic missile so that it avoids damage during the encounter with an ABM missile. It might be possible, for example, to improve the design of the warhead to this end. Appropriate shielding could make it better able to withstand blast and absorb neutrons or X-rays. Alternatively, it may be possible to choose materials used in the construction of the warhead which improve its penetration capabilities.

Another possible technique is to shoot a large number of decoys with the actual missile to saturate the radar system which form the 'eyes' of the ABM system. The booster stage of the missile could be made to fragment, after the separation of the warhead stage of the missiles, in such a way that the radars could not distinguish between the fragments and the actual warhead. Decoys are relatively simple to introduce once the attacking missile has left the earth's atmosphere. It would be feasible, for example, to arrange for the missile to eject balloons which automatically inflate in space. These would follow a similar trajectory in extra-atmospheric space and, if designed to be of an appropriate shape, would confuse the ABM radars until the missile re-entered the earth's atmosphere.

A large quantity of fine metal wires, usually called chaff, could be dispersed over a large portion of the sky by missiles. If the length of earth wire was equal to, or close to, the wavelength on which the ABM radars operated they could provide a large radar reflection which would conceal the incoming missile. Complete radar blackout, induced by a high level of ionization, could be produced by the explosion of nuclear warheads above the atmosphere. If offensive missiles were then fired during the blackout period they could escape detection by the ABM system. Jamming devices could also be used, either on the offensive missiles themselves or on separate missiles fired with the offensive missile, to confuse the ABM radar.

Another possible penetration aid is to design the offensive missile in such a way that it is capable of modifying its trajectory in mid-course or near the terminal section of its trajectory. The information used by the ABM to plot the trajectory would probably be based on the first part of the trajectory. The defensive missiles would then be fired according to this data. If the trajectory of the incoming missile changed direction after the ABM had received its instructions then the missile could avoid the interception. If the incoming missile is provided with several independently targetable warheads, which could be released at different stages of the missile trajectory, the ABM system would have to track and intercept each of these warheads separately. This would greatly increase the probability of one or more of the warheads getting to their targets.

Many of these penetration aids are most effective against the long-range interceptions used for area defence and some are only effective against area defences. This should not, however, be taken to mean that point defence is easier. Not only are warning times very much shorter, only a very few seconds in fact, but it is relatively easy for an attacker to by-pass defences by choosing an appropriate strategy. A number of strategic options are open to the attacker to overcome ABM defences. He could, for example, increase the number of his offensive missiles either directly or by providing each missile with several warheads. Alternatively, he could concentrate on the key radar installations of the ABM system by saturating the installations with a first blow and then follow this up by the actual missile attack on strategic targets. Attacking relatively small areas with many missiles would effectively saturate those parts of the ABM system defending these areas and increase the probability of a successful attack. It might be possible to attack an urban area defended by point ABM defences by targeting the offensive missiles at undefended areas upwind of the populated regions and designing the warhead to maximize fall-out, thus relying on radiation to produce high casualty rates in the population. Point defences could also be rendered useless if very large offensive weapons (say 100 megatons) were detonated at high altitudes above the ground. The ABMs would then be unable to reach the incoming missiles and the large nuclear explosion would destroy the target by the incendiary effects of the fireball produced by the explosion. It is, of course, possible to imagine other, more bizarre, methods of overcoming extensive ABM systems. Nuclear weapons could be introduced into enemy territory by hiding them in cargo ships or they could be detonated off the enemy coast by submarines to produce large tidal waves to destroy targets on or near the coast. It is possible that the deployment of ABM defences could cause a shift to other weapons of mass destruction, such as biological or chemical weapons.

## DEPLOYMENT OF NEW WEAPON SYSTEMS

### *MIRVs*

The development and deployment by the superpowers of multiple warhead missiles or MIRVs (Multiple Independently-targetable Re-entry Vehicles) is stimulated by the deployment of ABM systems. This development has major implications for the arms race, for disarmament and for its destabilizing effects mainly because it makes the calculation by each superpower of the size of the other's strategic forces impossible; the consequent erosion of the nuclear stability of the superpowers is clear. A new method of appraising the superpower's deterrents will therefore be necessary. It is not the number of missiles that is important but the types of warheads they carry. MIRV development also places a new premium on delivery vehicles with large thrusts and on invulnerable missile launchers, such as submarines.

Both superpowers have been developing a single ICBM able to carry several individually propelled warheads which can be guided to different targets hundreds of miles apart. For example, in the United States, the Minuteman-3, which is replacing the Minuteman-1, and the submarine-based Poseidon missile, which will replace the Polaris missile, will be equipped with MIRVs. The purpose of these missiles is to saturate the enemy ABM system. The United States Defense Department has revealed that the United States is developing a MIRV with a guidance system and thruster rockets so that it can make minor manœuvres after the main booster rocket has cut out. Travelling at an altitude of 800 to 1300 Km it could make a series of course and speed changes, ejecting a warhead each time at different targets. Because of the altitude these targets could be hundreds of kilometres apart and several degrees of longitude or latitude to either side of the trajectory. It is claimed that the vehicles carry up to 20 kiloton-size warheads. Each will have its own guidance system programmed to take it to a specific target.

### *Fractional Orbital Ballistic Systems*

The United States Defense Secretary announced earlier this year that the Soviet Union was probably developing a system for putting powerful space bombs in orbit around the earth. This system involves launching a nuclear warhead into a very low orbit, about 150 km above the earth (an ICBM may reach a peak altitude of 1300 km); at a given point, and before the completion of the first orbit, a retro-rocket would slow down the warhead, causing it to drop out of orbit on to the target. The potential warhead is estimated to be between one to three megatons. The orbit of a FOBS would have an almost constant altitude above the earth's surface whereas an ICBM follows a true ballistic trajectory which has much more curvature. Because of its low orbit a FOBS would not cross the radar horizon, and would therefore avoid detection by ground radar, until it was at a distance of about 1400 km. The warning time would be reduced to about 3 minutes compared with the corresponding warning times for an ICBM on a ballistic trajectory of about 10 minutes. Also, a missile on a fractional orbit trajectory could be made to approach the opponent's radar system from a direction which would ensure that it would not be detected. Modern phase-array radars operate in a fixed direction and over an angle of approximately 90°.

In connection with these and similar weapons Article 4 of the Outer Space Treaty approved by the United Nations in December 1966, should be noted. This states that 'States parties to the treaty undertake not to place in orbit around the earth any objects carrying nuclear weapons or any other kinds of weapons of mass destruction, instal such weapons on celestial bodies or station such weapons in outer space in any other manner.'

### *Future ABM Technology*

It has been announced that a new high-acceleration interceptor missile (about 500 g) is being developed by the United States for point-defence of ICBM sites. This project is known as the High-acceleration Booster Experiment, or HIBEX. Seaborne ABMs are also under development, known as Seaborne Anti-ballistic Missile Intercept Systems (SABMIS). These are presumably aimed at the interception of enemy ICBMs in the mid-course of their trajectories. Research and development to increase the range of the Spartan missile has the same purpose. Also space-based systems, to intercept ICBMs in the intial phase of their trajectory, are no doubt being actively considered. The United States Ballistic Missile Boost Intercept (BAMBI) concept, for example, envisaged the use of about 1200 orbiting satellites armed with intercept missiles. It was intended to cover the air space of the Soviet Union with enough ABMs to destroy ICBMs during their boost phase. BAMBI has not, so far, been developed, because of the high cost and low effectiveness of such a system.

The ABM systems in existence are based on the principle of the 'bullet-to-bullet' duel defence in which interceptor missiles are fired at hostile strategic missiles which have been aimed at specific targets. Other systems are, however, feasible. For example, it has been suggested that an effective defence screen could be established by high-megatonnage warhead explosions set off in space by defensive missiles, inside the earth's magnetic field. The charged particles released by these nuclear explosions would move along the lines of the earth's magnetic field and may achieve a density sufficient to inactivate an incoming enemy warhead. The problem would be that the particles would leak from the magnetic belt too fast to maintain the required flux. Nevertheless, there have been speculations that the Soviet Union is developing some kind of 'shield' anti-missile system. Alternatives to a plasma of charged particles that have been suggested for defence screens include small pellets and gases. The future use of lasers for ABM systems is another possibility if it becomes feasible to project sufficiently large fluxes of radiation over great distances. Methods of destroying an opponent's offensive missiles during the very early phase of their trajectories, for example during the boost phase, have been suggested. During this period it would be extremely difficult for the attacker to take counter measures or counter strategies to the ABM system. It has been suggested in this connection that ABMs could be installed on the sea-bed close to the enemy's shores.

### THE PRESENT STRATEGIC BALANCE

The comparative strategic strengths of the nuclear powers, as estimated by the Institute of Strategic Studies, for early 1969, are shown in Table 5. This balance is important to the discussions on the effects of the deployment of ABM systems which follow in later chapters. The major nuclear missile delivery systems of the superpowers are shown in Table 6.

In the United States Minuteman-2 solid-fuel ICBMs are replacing the Minuteman-1 and there are already about 350 Minuteman-2 missiles deployed. The Minuteman missiles are installed in hardened silos. An improved ICBM, Minuteman-3, is under development which will be equipped with MIRVs. Some of these missiles have already been tested. Apart from the Minuteman missiles the United States has about 50 Titan-2 liquid fuelled ICBMs remaining in service. Each of the 41 nuclear ballistic missile submarines carries 16 Polaris missiles. Eventually it is planned to replace the Polaris missiles with the Poseidon missile, and the production of this missile, which will be equipped with MIRVs, has begun. The American strategic forces could deliver at least 2000 thermonuclear weapons, with an average yield of one megaton.

The number of ICBMs deployed by the Soviet Union has nearly doubled in 1968 and ICBMs with multiple warheads have been tested. About 50 Soviet submarines can fire ballistic missiles and each submarine carries, on average, three missiles. The Soviet navy is bringing into service a new class of nuclear ballistic missile submarine similar to the American Polaris and capable of carrying 16 missiles. China has probably produced enough fissile material for nearly 100 nuclear weapons of nominal (20 kiloton) yield and for some thermonuclear weapons. It is likely that China will soon deploy short- and medium-range missiles and is almost certainly developing ICBMs, although not at such a fast rate as some Western observers have forecast.

TABLE 5

*Estimated Comparative Strategic Strengths of the Nuclear Powers*

| *Category* | *China* | *France* | *U.K.* | *U.S.A.* | *U.S.S.R.* |
|---|---|---|---|---|---|
| *Missile and air power* | | | | | |
| Land-based ICBMs | – | – | – | 1054 | 900–1000 |
| Fleet ballistic missiles | – | – | – | 656 | 125 |
| IRBMs and MRBMs | * | – | – | – | 750 |
| Long-range heavy bombers | – | – | – | 500 | 150 |
| Medium bombers | 10 | 50 | 75 | 40 | 1050 |
| *Sea Power* | | | | | |
| Aircraft carriers | – | 3 | 2 | 23 | – |
| Helicopter carriers, commando carriers and assault ships | – | 3 | 4 | 160 | 1 |
| Guided missile cruisers | – | – | – | 12 | 7 |
| Gun cruisers | – | 2 | – | 2 | 12 |
| Large guided missile destroyers and frigates | – | 2 | 6 | 55 | 28 |
| Ocean-going escorts | 18 | 44 | 66 | 267 | 178 |
| Nuclear-powered missile submarines | – | – | 1 | 41 | 38 |
| Other missile-firing submarines | 1 | – | – | – | 50 |
| Nuclear-powered fleet submarines | – | – | 2 | 35 | 12 |
| Other submarines | 32 | 19 | 30 | 78 | 280 |

* China is expected to have IRBMs in 1969 and some ICBMs by the early 1970s. ICBM – Intercontinental ballistic missile; IRBM – intermediate range ballistic missile; MRBM – medium range ballistic missile.

TABLE 6

*Major Nuclear Delivery Systems (missiles) of the Superpowers*

| | *Name* | *Range (miles)* | *Estimated Warhead* |
|---|---|---|---|
| U.S.A. | LGM-25C Titan 2 | 9,000 | 5 + megaton |
| | LGM-30A Minuteman 1 | 6,500 | 1 + megaton |
| | LGM-30F Minuteman 2 | 7,900 | 2 megaton |
| | UGM-27A Polaris A1 | 1,380 | 0·7 megaton |
| | UGM-27B Polaris A2 | 1,700 | 0·7 megaton |
| | UGM-27C Polaris A3 | 2,850 | 0·7 megaton |
| | MGM-13B Mace | 1,380 | kiloton range |
| | MGM-31A Pershing | 400 | kiloton range |
| | MGM-29A Sergeant | 75 | kiloton range |
| U.S.S.R. | ICBM Scrag | Orbital | ?30 megaton |
| | ICBM* | 5,000 + | 10 megaton |
| | ICBM* | 5,000 | 10 megaton |
| | ICBM Sasin | 5,000 | 5 megaton |
| | ICBM* | 10,000 | 20 megaton |
| | ICBM Savage | 6,000 | 1 megaton |
| | IRBM Skean | 2,100 | 1 megaton |
| | MRBM Sandal | 1,100 | 1 megaton |
| | SLBM Sark | 400 | 1 megaton |
| | SLBM Serb | 650 | 1 megaton |
| | SLCM* | 300 | kiloton range |
| | SLM Strela | 400 | kiloton range |
| | SRM Scud | 70 | kiloton range |
| | CRM Shaddock | 250 | kiloton range |

* No name assigned: LGM – silo-launched missile; UGM – underwater-launched missile; MGM – mobile-guided missile; SLBM – submarine-launched ballistic missile; SLCM – submarine-launched cruise missile; SRM – short-range missile; SLM – ship-launched missile; CRM – cruise missile.

C. F. BARNABY

# 2: Arguments For and Against the Deployment of Anti-Ballistic Missile Systems

*Introduction*

Many arguments have been advanced both for and against the development and deployment of ABM systems (Table 7) and it might be useful to summarize the principal ones here. It must be emphasized, however, that some of these arguments are based on very sophisticated premises and it has been necessary to simplify them in the following discussion.

The present strategic situation between the superpowers is based on mutual deterrence which is assumed to guarantee stability on the strategic level. Deterrence is provided by the so-called second strike assured destruction capability of each superpower. This is the capability to inflict an unacceptable degree of damage on the other side by a retaliatory strike even after absorbing an all-out surprise first strike, even though the latter is aimed at strategic forces. There is, of course, a psychological element in the determination of what the adversary would regard as 'unacceptable' damage. Because strategic stability is based on the effectiveness of the threat of assured destruction to deter either side from launching a surprise attack the preservation of the capability of second strike assured destruction is regarded as of fundamental importance by the decision-makers in both superpowers.

A categorization of new weapons is possible in terms of their effect on the strategic equilibrium between the superpowers based on deterrence. New weapons which reinforce the second strike capability against enemy cities and populations can be regarded as stabilizing whereas new weapons which increase first strike capability against the enemy's strategic forces are destabilizing. This categorization arises because the enemy's cities and populations are the hostages when deterrence is based on a balance of terror. The theory suggests than an aggressive first strike by strategic weapons would be aimed at the enemy strategic forces (i.e. would be a counterforce strike). Second strike assured destruction, and strategic balance, requires that sufficient strategic forces would remain intact after such a first strike to enable the superpower who has absorbed it to inflict a level of damage on the other side's cities and population which this superpower, who made the first strike, would regard as so high as to deter him from making the strike in the first place; a strike against cities and populations is called a counter-value strike. Thus, nuclear deterrent strategy differs from conventional strategy in that weapons aimed at populations are considered, in a sense, less dangerous than weapons directed at the enemy's strategic forces. A similar difference arises with defensive measures since those designed to preserve strategic forces are stabilizing whereas those designed to protect populations are destabilizing to the extent that they reduce the effectiveness of the other side's retaliatory second strike counter value capability. In practice, if one side deploys destabilizing weapons it is likely to lead to a reinvigoration of the arms race by causing the other side to react, usually by increasing the number or performance, or both, of his strategic forces in an attempt to restore the strategic deterrence balance. The effects of the development and deployment of ABM systems should be examined within this framework.

### *Arguments for ABM systems*

An argument often given for ABM systems is simply based on the feeling that all defensive measures are desirable and that if ABM systems, should war occur, save some lives their deployment will have been justified. This is an unsophisticated and emotional argument which has political appeal but little relevance in the complex concepts of deterrence. It is, however, a perfectly understandable argument in countries, like the Soviet

Union, that have suffered horrendous damage and casualties in past wars and naturally carries great weight with populations that have vivid memories of such experiences.

A similarly irrational argument for an ABM system is that it should be developed and deployed because the other side has developed or deployed one. This also has psychological and political appeal but ignores the fact that the logical response of an opponent to an ABM system is to increase the number and/or the performance of his offensive forces.

A much more powerful argument for ABM systems has been put forward by serious students of the subject. This is based on the theory that if ABM systems are deployed extensively by both superpowers and if their offensive forces are not increased, or better still if they are decreased, then the strategic equilibrium between the superpowers will no longer be based on deterrence but, instead, will be based on defence. The advocates of this theory maintain that the situation after the transition would be less dangerous, in that general nuclear war would be less likely, and that nuclear disarmament would be much easier to negotiate. If offensive forces were reduced the deployment of ABMs would, of course, amount in itself to a measure of disarmament. The theory requires that the superpowers mutually agree to undergo the necessary changes in their strategic and defensive postures. There are several difficulties with this argument. It can be shown that the transition from deterrence to defence involves passing through a period of great strategic instability between the superpowers during which the chance of general nuclear war is greatly increased compared with the relatively small chance of such a war which exists with the present strategic balance. Furthermore, the amount of mutual trust which would have to exist between the superpowers to achieve the transition is unlikely to occur in the foreseeable future, and, in fact, if such trust between them did exist then a great deal of disarmament would, in any case, be readily negotiated. In such a world high levels of strategic or defensive forces would be irrelevant. Experience has shown that the most likely result of the deployment of ABMs by one side is an increase, rather than a decrease, in the quantity or quality of the offensive forces of the other side and this is likely to continue to be true for the foreseeable future.

Some people see in the deployment of an ABM system a possibility of reaching a dominant or 'winning' posture and this is used as a reason for acquiring the system, usually by the more militant groups in the society. Such groups find the present situation of stability enforced by deterrence to be a frustrating situation and this probably accounts for the pressures from the military establishments within the superpowers for the acquisition of large ABM systems. These pressures are, at certain times, powerful and should not be underestimated.

The official argument for the deployment of the United States ABM system is to protect against the possibility of an attack against China at the time when she has deployed a small strategic ICBM force, generally thought to be the early or mid-1970s; such an attack would be irrational since it would be suicidal. It is also claimed that the system would eliminate the possibility of catalytic war initiated by a small nuclear power and would insure against accidental war caused by the misfiring of a missile by any one of the nuclear powers. To achieve this insurance the system would have to be fully alert continuously and automatic operation would be necessary since warning times could be extremely short. It could be argued that the chance of accidents associated with the ABM system itself, operated under these conditions, is greater than the chance of the occurrence it is insuring against. It has also been argued that a 'thin' ABM system would prevent, or reduce, the credibility of small token attacks by ICBMs made for bargaining purposes in times of crisis.

It has been claimed that ABM systems discourage the proliferation of nuclear weapons by adding credibility to the guarantees given or offered by superpowers to allies. It is also claimed that the deployment of complex weapons systems by the superpowers increases the gap between them and other powers. Because a potential new nuclear weapon power could not hope to compete with such sophisticated weapons it is argued that they are unlikely to start along the path. Although this argument might apply if these powers were intending to challenge, or deter, a superpower it does not apply if the perceived opponent is another small power.

Because ABM systems can be used to strengthen second strike capabilities it is argued that they can be stabilizing. In one sense this argument is true, but because an ABM system can be overcome if the other side increases his offensive weapons the deployment of them is likely to lead to an acceleration of the arms race. It should be noted that neither side, so far as is known, is using ABMs to protect offensive missiles. Such a role would, in fact, be inconsistent with the declared anti-Chinese purpose of the United States Sentinel system.

*Arguments against ABM systems*

A main argument against an ABM system is that it is not, and is not likely to become, effective. This means that ABM defence cannot be made adequate relative to the offence existing at the time the system becomes operational. There are some, for example, who doubt whether the Sentinel system will be effective even against attack by the first generation of Chinese ICBMs. An ABM system is a complex system of many components and, since none has been fully tested under operational conditions, it is not possible to assess the effectiveness of ABMs. However, it is possible to predict with some confidence that an ABM system can be relatively easily overcome by the use of penetration aids or suitable strategies. There seems to be little doubt that at the moment the cost-exchange ratio favours the offence rather than the defence for all except very high levels of casualties and damage.

Another main argument against ABM systems is that they have resulted in a reinvigoration of the arms race. This arises because the likely response to a deployment of an ABM system by superpower A is for superpower B to increase the performance and/or numbers of her offensive weapons. This increase is likely to lead to the deployment of a heavier ABM system by superpower A in an attempt to protect against the increased strategic forces of superpower B. Superpower B would then probably respond by a further increase in offensive forces and so on. A second effect on the arms race arises from the uncertainty of the effectiveness of ABM systems. If a superpower deploys an ABM system the other superpower is likely to assume that the effectiveness of it is greater than it actually is and therefore to overreact to the deployment. On the other hand, the first superpower is likely to assume that the effectiveness of his ABM system is less than it actually is and to employ a heavier system than is necessary for given defence requirements. Moreover, if one side begins the deployment of an ABM system the other will probably tend to respond, not to the level of the deployment, but to a level which this superpower calculates could be deployed by the other if it devoted a maximum effort to the deployment.

Other arguments against ABMs arise from the possible effect of these weapons on the arms race. For example, there is the effect on the *détente* between the superpowers. During a period of uncertainty concerning the performance of offensive and defensive weapons an atmosphere of fear and distrust is likely to arise which is not conducive to a *détente*. This is particularly true if the weapons developments are such as to cause the superpowers to perceive a possibility of attaining a dominant position in the long run. The situation then changes from one based on coexistence to one based on competition for domination. In practice, neither superpower is likely to attain a winning posture but the perception of the possibility might be sufficient to produce pressures within superpowers to attempt to attain it.

The decision to deploy an ABM system, even a 'thin' system, is regarded by some as the thin end of the wedge which will almost inevitably be followed by the deployment of a heavier system. Experience has shown that it is undoubtedly difficult to keep any weapons system limited. Once the deployment has begun powerful pressures build up for an extension of the system. As we have seen, this is particularly likely in the case of ABM systems since the logical response to the deployment, namely an increase in offensive strength, is conducive to pressures for a heavier system to counter the increased offence. Because ABM systems could stimulate a vicious circle of increases in weapons it is argued that the process should not be started. Political pressures would be greatest if ABMs were used for the point defence of cities. It would be extremely difficult to limit such a defensive system since if some cities were selected to be defended then the populations in other cities would strive for equal protection. It is hard to imagine that the United States Administration, for example, could limit ABM point defence to only a few cities.

A powerful argument against ABM systems is the probable effects of their deployment on the prospects for gradual disarmament and arms control. Most authorities agree that successful arms control negotiations are probably only possible during a period of strategic stability and *détente* between the superpowers. Some arms control measures have, in fact, been negotiated since 1963, a period during which both superpowers perceived that a strategic balance had been reached. It is when both sides feel relatively secure that they are most likely to consider means to limit the arms race, an objective which is in the interests of the superpowers so that they can concentrate on their internal problems. For these reasons, any development which erodes or destroys the present strategic balance between the superpowers might also destroy any prospects there may be for the negotiation of arms control and disarmament measures. We have seen that the deployment of an ABM system by one side is likely to cause the other side to develop and deploy countermeasures, including MIRVs. An analysis of the effect of ABMs and MIRVs on the strategic balance shows that the extensive deployment of these two weapon systems is particularly destabilizing, particularly if these weapons are made accurate and effective. Although the presently planned deployments of ABMs do not seriously impair the strategic balance between the superpowers it is argued that if the development of new weapons is allowed to continue the strategic balance will be endangered and therefore restraint should be used in the development and deployment of these weapons. It is important to stress that it is not argued that the present strategic balance is to be maintained at all costs because it is a particularly desirable state of affairs but rather that under the present condition of International Society disarmament and arms control can be negotiated only if the superpowers feel secure. The maintenance of the strategic balance is merely a convenient method of providing this necessary feeling of security. Nor is it suggested that the present levels of armaments were arrived at as a deliberate policy to provide a balance but rather that the superpowers have reached their present situation as a result of the uncontrolled momentum of the arms race. The argument is that having arrived at the present situation and having realized the possibility of negotiating arms control measures from the present levels of arms the superpowers should not allow the opportunity to be jeopardized by the development and deployment of new arms, such as ABMs, and the consequent acceleration of the arms race. This amounts to a plea for deliberately controlling the momentum of the arms race. Those that advance these arguments also point out that this might be the last opportunity for the superpowers to negotiate gradual disarmament and are therefore particularly anxious to prevent the deployment of extensive ABM systems.

The deployment of extensive ABM systems is likely to have adverse effects on the possibility of the negotiation of a comprehensive test ban treaty and also on the viability of the Non-proliferation Treaty. It is unlikely that a heavy ABM system would be deployed unless the system was, to some extent, tested under realistic conditions and, therefore, even the partial test ban treaty would be jeopardized. It is unlikely that the Non-proliferation Treaty will be viable unless the superpowers take seriously their obligations under Article 6 of the Treaty which commits them to take . . . 'effective measures relating to cessation of the nuclear arms race at an early date and to nuclear disarmament . . .'. If the superpowers deploy extended ABM systems then it will seriously put into question their good faith towards their treaty obligations.

### *Conclusions*

Most authorities feel that the arguments against the deployment of ABM systems completely outweigh the arguments for such deployment. The question then arises of why, in spite of this, the superpowers have decided to deploy ABMs. So far as the Soviet Union is concerned it is probable that the traditional role of defence in Soviet military planning played an important part in the decision. Furthermore, the Soviet people, who suffered unimaginable losses in World War II, would clearly favour any defensive measure. This would make the task of those within the decision-making powers who favoured deployment easier, since those who attempted to resist deployment could be accused of arguing that the country should be left undefended. The fact that the Soviet ABM system is not very extensive may be taken to indicate that although the deployment was initiated it has not been actively pursued. It is known that some Soviet experts are against the deployment of ABMs.

There is no doubt that political pressures within the United States for the deployment of ABMs became strong after the Soviet Union had deployed their system. It is likely that Sentinel was deployed mainly for political reasons. The political capital made out of the supposed missile gap during the Kennedy election campaign demonstrated the dangers of appearing to be dilatory about strategic and defensive matters, a state of affairs that must make political leaders sensitive to such pressures. The supposed threat from China probably also played an important part in the decision although arguments based on this reason cannot really be substantiated. It is likely that pressures from the so-called military-industrial complex was also a determining factor. As we have seen, these groups are in favour of the deployment of a 'thin' system as a first step towards the deployment of a heavier system at some future date. Experience has shown that when the administration is not prepared fully to concede their demands, a policy of settling for a lesser system is adopted and then, once this has been installed, pressure for a larger system is brought to bear. An ABM system is so complex that very large investments in the electronics, aerospace and other industries are involved. It is, therefore, perhaps understandable that those with vested interests in these industries should be in favour of the deployment of an ABM system. It is well known that these sectors have powerful lobbies in the American political system.

Finally, some authorities argue that the momentum of technology is irresistible and, therefore, that the continued development and deployment of new weapons systems is an almost inevitable process. These people would regard the Soviet and American ABM systems as part of this process.

TABLE 7

*Summary of Arguments For and Against the Deployment of an ABM System*

| *Arguments for* | *Arguments against* |
|---|---|
| Defensive measures are good by definition. | ABMs are ineffective, the cost-exchange ratio favours the offence. |
| The other side has it therefore we must. | It reduces the possibility of a comprehensive test ban treaty. |
| It will prevent a catalytic, nuclear war provoked by small nuclear powers. | It threatens the partial test ban treaty. |
| It will insure against nuclear war started by the accidental launching of one or more ICBMs. | It demonstrates that Man cannot control the technological revolution. |
| The transition from equilibrium based on deterrence to equilibrium based on defence would produce a safer world and could itself be a disarmament measure. | It leads to a reinvigoration of the arms race and stimulates the deployment of very destabilizing weapons. |
| It might produce a 'winning' position. | It is likely to erode and destroy the strategic balance between the superpowers and the *détente*. |
| It is an insurance against deliberate attack by smaller nuclear powers (e.g. China). | It will be difficult to keep a 'thin' system limited. |
| It discourages the proliferation of nuclear weapons by widening the gap between the superpowers and the others. | It threatens the viability of the non-proliferation treaty and encourages proliferation of nuclear weapons. |
| It strengthens second strike capability. | It is likely to lead to a situation where disarmament negotiations are very much more difficult, if not impossible. |
| It will provide economic benefits in terms of employment, large investments, etc. | It is extremely costly, the money would be better spent in solving internal problems. |
| It reduces the possibility of smaller nuclear powers (e.g. China) using nuclear blackmail against allies. | It enhances, rather than detracts from, the nuclear deterrent of smaller powers. |
| It increases the credibility of guarantees to allies. | It creates uncertainty concerning the effectiveness of strategic forces. |
| It reduces the credibility of small token or 'bargaining' strikes in crises. | |

# Part II

# PAPERS PRESENTED TO THE SYMPOSIUM

## SECTION A: *Implications for the Arms Race*

N. PETERSEN

# 1: The Arms Race Implications of Anti-Ballistic Missile Defences

In discussions of the implications of ABM deployment, which have gained momentum during the past three or four years, the effects on the arms race have received particular attention. This has especially been the case in considerations of 'thick' city or population defences. It is often asserted that a main consequence of the deployment of such systems would be an acceleration or renewal of the strategic arms race between the United States and the Soviet Union, which would have deleterious effects on political relations between them and might ultimately lead to a new 'cold war'. Thus J. B. Wiesner argues that the deployment of ABMs would lead to 'a frantic new round of defence spending' resulting in greatly increased levels of damage capacity on both sides because of an almost certain tendency for both parties to over-rate the effectiveness of each other's defensive systems and therefore to build considerably more offensive power than would actually be required.[1] A less passionate, but equally pessimistic view has been taken by the Johnson Administration whose exspokesman on these matters, Secretary of Defense Robert S. McNamara, said in his 1967 Posture Statement:

> . . . all we would accomplish (by deploying Nike-X for population defense against a Soviet attack) would be to greatly increase both their defense expenditures and ours without any gain in real security to either side.[2]

The Soviet position, on the other hand, has been – at least until recently – to play down (but not to deny) the arms race implications of ABMs and to point to the positive aspects of ABM systems, primarily the fact that they are nominally defensive, rather than offensive, weapons systems.[3]

Before considering the substance of these and other statements on ABMs and the arms race, it may be useful to have a more general discussion of the arms race problem as it relates to ABMs.

By 'arms race' I shall refer to an interaction between the Soviet and the American weapons programmes which meets two criteria:

1. actual force levels (*in casu* strategic force levels) must increase on both sides;
2. the main stimuli for these increases must be the weapons programme of the opposing power or, in other words, both sides must be motivated by a determination to improve, or to keep equal, their relative strategic strength.

This definition is trivial. On the other hand, it is often forgotten that other motivations, which are not directly related to political and military competition, may play a major role in the formulation of weapons programmes; e.g. inertial conversion of Research and Development projects into military hardware (which seems to be a factor of increasing rather than decreasing importance, ABMs being no exception), a wish to pacify the military establishment, industrial pressure, etc. In fact, differences in motivation – or rather motivation as perceived by the opposing side – may be of major importance in deciding whether a particular move will or will not provoke an arms race. It may be argued that the less A perceives a move of B to be motivated by primarily competitive considerations, the less likely he will be to respond in kind. To take a topical example: those in the United States who view the present Soviet ABM deployment as motivated by a traditional Soviet preoccupation with defence seem less inclined to think that it calls for determined counter-measures by the United States than those who believe that it is inherent in a Soviet effort to reverse the strategic balance and to gain strategic superiority *vis-à-vis* the United States.

What is important, then, are perceived, rather than 'real',

motivations for increasing a nation's strategic force level. Such perceptions will be conditioned by several factors, among them the prevailing 'climate' in international politics. Perceptions will generally be more alarmist in a period of high international tension than in more tension-free ones, which suggests that an ABM deployment in the present period of, at least, limited *détente* may be less likely to create arms race interactions than would have been the case some years ago. Another determining factor is the mode of strategic thinking of the two parties. It seems reasonable to think that the United States, which – in theory at least – is committed to a rather pure deterrence philosophy, will be more alarmed by a massive Soviet ABM deployment and more likely to respond with offensive force build-ups than would the Soviet Union with its more 'traditional' strategic thinking in the case of a corresponding deployment by the United States. Conversely, the Soviet Union would presumably be more alarmed by a major offensive force build-up by the United States, e.g. the introduction of MIRVs in a numerically stable missile force, than decision-makers in the United States would be. This basic incongruity between Soviet and American defence thinking should be seen as an important factor in determining whether ABM systems can be introduced without releasing vigorous arms race stimuli. Generally speaking, the arms race implications of an ABM deployment seem more serious if the strategic outlooks continue to diverge rather than converge.

Reactions to ABMs would also, to a great extent, be conditioned by beliefs about its 'objectively' destabilizing effects. Both powers would certainly react strongly, if an ABM deployment were seen as strengthening the opponent's first strike capability. If so, a dangerous situation of preemptive instability would have been created. In the American ABM literature it is often asserted that ABMs are destabilizing in this sense.[4] The argument is that missile defences work better against a second strike (which would be rugged and un-coordinated, and against which they would be fully alerted) than against a well-coordinated first strike which would have the benefit of surprise. This argument is plausible as far as it goes. It is less clear, however, how great the difference is, and whether it is great enough to have any significant bearing upon the choice between peace and war. If we presuppose that both sides have missile defences, then a first strike attack by A would be degraded to some extent by B's defences which would mean that B's second strike would be degraded, too, by A's defences, but so would A's second strike. Which side would be the 'loser' as compared to a situation in which there are no ABMs, may thus be quite difficult to predict, and given the present odds against first strike strategies any differences may be of marginal importance.

The situation might, however, be different in a strategic environment which increased the feasibility of counterforce strategies. Such a situation might, for example, result from the introduction of MIRVs (which will probably be deployed irrespective of developments in the defensive field) which would tend to change significantly the attacker/target ratio in favour of counter-force attacks. This argument of course, presupposes that MIRVs can be given a counterforce capability. However, according to the U.S. Defense Department, MIRVs 'will be far more accurate than any previous or existing warhead. They will be far better suited for the destruction of hardened enemy missile sites than any existing missile warheads'.[5] The main feature of such a development would be a marked reduction of the numerical superiority (in terms of missiles) required for a successful counterforce attack. These prospects have as yet remained relatively unexplored[6] in comparison to the very great interest which has been accorded to the possible destabilizing effects of ABMs. In fact, if the presupposed situation comes to pass, ABMs (especially hard-point defences) may plausibly be seen as a restabilizing factor by virtue of their reduction of the attacker/target ratio to more secure levels. On balance, therefore, it is not obvious that ABMs should be dangerous first strike weapons and on that account necessarily release vigorous arms race stimuli. The best answer to the MIRV problem would undoubtedly be for both superpowers to do away with land-based missiles entirely and to concentrate on submarine-based missiles. Service rivalries, control problems and a general wish to avoid putting all eggs in one basket would, however, militate against this notion.

On the other hand, the high degree of uncertainty associated with all evaluations of ABM problems would tend to strengthen rather than weaken the arms race stimuli of an ABM deployment. The inclusion of ABMs would complicate the present strategic balance (a tendency towards greater complexity which would be further underscored by the introduction of MIRVs) and might thus lend itself to propagandistic manipulation by one or both parties. Experience from the 'missile gap' period illustrates this danger (even though the same experience shows that such manipulations of the strategic balance can be counter-productive in its effects). In any case, it can be predicted that estimates of the strategic balance will be more conservative in an ABM world, and that the tendency of decision-makers to hedge against 'greater-than-expected' threats will increase. It should be mentioned, too, that the great uncertainty as to the evaluation of the effectiveness of future defensive and offensive weapons systems will probably increase the possibility of 'internal' arms races developing within the respective American and Soviet weapons programmes.

The remaining part of this chapter will concentrate on the conditions under which an arms race stimulated by ABM deployment may start, how it may develop, and how it can be controlled so that destabilizing effects are minimized. The arguments adduced will rest on three assumptions, viz.

1. that both the United States and the Soviet Union will react to what they perceive as unfavourable changes in the strategic balance,
2. that their capability to do so is about equal (this is less certain), and
3. that there will be no formal or tacit agreement between them to 'outlaw' ABM systems. (In view of the recent announcement that talks on the limitation and reduction of offensive and defensive strategic weapons systems are going to start in the near future, this assumption is arbitrary. For many reasons, which cannot be stated

here, it seems at least an open question, however, whether these talks will result in some sort of agreement to prohibit the deployment of ABMs.)

In theory, the fact that offence and defence (assured destruction and damage limitation respectively) are two sides of the same coin makes three patterns of arms race interactions possible:

1. an offensive-defensive pattern, which may be divided into two categories:
   (*a*) a 'one-dimensional' pattern in which one part stresses its defences while the opponent seeks to preserve the status quo in terms of its assured destruction capability by further offensive force build-ups; and
   (*b*) a 'two-dimensional' pattern in which the offences and defences of both parties compete with each other.
2. an essentially defensive-defensive pattern, i.e. both parties give their ABM efforts priority over investments in offensive weapons systems, and
3. an essentially offensive-offensive pattern which will be fairly familiar from the missile race.

What particular patterns will emerge, and how dynamic interactions between the respective weapons programmes will be, probably depends to a considerable extent on how interception and penetration technology develops in the coming years, and thus on the so-called 'cost-exchange ratio' (the ratio of ABM costs to offsetting offence costs). Generally, one would expect patterns 1 (*a*) or 3 to be typical of situations in which defences are relatively uncompetitive, and patterns 1 (*b*) or 2 to be typical of situations in which missile defences compete with offensive systems on a more equal footing. McNamara gave the version on p. 45 of the present (or rather early 1967) relationship in his 1967 Posture Statement.[7]

It should be noted that these calculations are beset with many uncertainties. However, they illustrate the important fact that missile defences are cheaper and more effective at the upper end of the 'casualty scale' than at the lower end. They should

| *Levels of U.S. fatalities which the Soviet Union may consider to provide deterrent* | *Cost to the Soviet Union of offsetting U.S. cost to deploy an* ABM |
|---|---|
| 40 million | $ 1 Sov. cost to $ 4 U.S. cost |
| 60 million | $ 1 Sov. cost to $ 2 U.S. cost |
| 90 million | $ 1 Sov. cost to $ 1 U.S. cost |

give, too, a rough picture of the present relationship between offensive and defensive missile techniques. How these relationships will develop in the future cannot be predicted with any degree of certainty. However, the cost-exchange ratio seems to have changed considerably in favour of ABMs since only a few years ago it was estimated to be of the order of 10 : 1 or even 100 : 1.[8] It thus seems reasonable to expect the trend towards greater competitiveness for ABMs to continue. For the sake of argument the author will consider the arms race implications of two situations: the one referred to in McNamara's estimate, and some hypothetical future situation in which the cost-exchange ratio between ABMs and offensive weapons systems has changed by a factor of, say, 5 in favour of ABMs. (If the trend were to go in the other direction, the problems discussed here would lose much of their urgency.)

In the first case, ABM deployment would be economically competitive only within those very high ranges of the 'casualty scale' (above 90 million) where differences of say 10 million casualties in the expected outcome of a strategic nuclear exchange would be of practically no significance either for deterrence or for war-fighting purposes. The effects of the present limited Soviet ABM deployment and of the American Sentinel system (which will also have some effectiveness against the Soviet Union),[9] lie probably at this upper end of the 'casualty scale'. The Soviet deployment of a Galosh system around Moscow has already given rise to an incipient arms race since the decision of the United States to accelerate the deployment of new, sophisticated offensive weapons systems like Poseidon and Minuteman-3 must be seen as a reaction to it. Likewise, the decision of the United States to deploy the Sentinel has been followed by Soviet references to the continued penetration capability of Soviet missile forces.[10] What is important, is that arms race interactions are unlikely to be very dynamic and destabilizing at the levels of mutual damage capability discussed here. The outcome may be that damage capabilities increase or decrease somewhat, but differences will probably be relatively small. Both superpowers will continue to have at their disposal, missile forces that will be able to inflict damages on the other power in a second strike on a scale that would serve as a reliable deterrent of any action that can be deterred.

Present cost-exchange ratios, of course, are not a guarantee against massive deployment of ABMs; on the other hand, it is difficult to believe that any country would try, over an extended period, to reduce its expected casualty rate in a strategic exchange (i.e. the opponent's assured destruction capability) well beyond the point where cost-exchange ratios begin to favour the offence, against an opponent who is determined to preserve his relative military strength. If such efforts were made, the opponent would find it progressively easier and cheaper to nullify them and so to fend off changes in the strategic balance. And, since 'over-insurance' by offensive force build-ups would be relatively cheap, the ABM power might in the end find that its relative strength had been reduced. Thus, McNamara has asserted that 'as a result of our offsetting actions the Soviets will be worse off than if they had never started to deploy an ABM system in the first place'.[11] This would probably mean that, even if an attempt were made against all odds, it would not be of long duration.

A different situation would arise if the cost-exchange ratios between defensive and offensive weapons systems were to change in favour of defence by a factor of 5. In that case the cost-exchange ratio would be about unity for those parts of the respective 'casualty scales' where each party might begin to feel worried about its continued deterrence capability. McNamara has estimated[12] that the capability of the United States to destroy one-fifth to one-fourth of the Soviet population and one-half of her industrial capacity would serve as an effective deterrent. The corresponding Soviet requirement is unknown, and it is, indeed, unknown whether the Soviet leaders believe that some fixed destruction capability constitutes the deterrent while other levels are considered either as insufficient or as a waste of resources.

In the hypothesized situation, which is still at least some years ahead, each power would, within the indicated limits, be economically free to choose between reducing its own casualties in a nuclear exchange by a certain amount or raising the casualty level of the other power by the same amount. Barring an agreement between the superpowers not to deploy ABM systems it seems reasonable to believe that both the United States and the Soviet Union would then try to buy security through damage limitation. In any case, if only one of them tried, the other would probably have to do the same, because it could not afford to lag behind in an important technological development. Theoretically, the other power might try to offset the opponent's ABM system by an offensive force build-up, but to restore the *status quo ante* would be a costly affair.

Less clear, however, is what the actual interaction between the two parties' offences and defences would be; i.e. whether a 'two-dimensional' offensive-defensive arms race pattern or an essentially defensive-defensive pattern would result. It is easy to visualize a situation of the first kind. It would typically occur if both powers believed they could reduce the other's damage capability to some low level while at the same time trying to keep their own damage capability intact. The efforts of the two powers would then be mutually exclusive, and the interaction between them would resemble that of a 'zero-sum game'. The result would, in all probability, be a vigorous arms race which would be further exacerbated by the general uncertainty prevailing in a strategic situation of such complexity. In theory, the impasse characteristic of a situation in which neither offence nor defence can gain the upper hand, can only be resolved in two ways; either through the financial exhaustion of the weaker power or by a change in the motivations of the two powers. The second solution seems much more plausible than the first. In practice it would mean that both settled for something less than the desired situation (maximizing both damage limitation and assured destruction) and accepted some sort of balance between their respective assured destruction and damage limitation capabilities. The focal point (an important concept in all bargaining whatever the idiom may be) for such a balance would be the point where the cost-exchange ratio of ABMs to offensive weapons systems is about unity. In concrete terms, such a balancing of the arms race would mean that investments which are uncompetitive – offensive investments above the balancing point and defensive investments below the point – would be avoided. A definite stabilization of the arms race cannot, of course, be expected, if only because cost-exchange ratios would change continually and because the calculations of the two powers, and of their respective military establishment, would differ (Soviet calculations of the effectiveness of ABMs seems to be more optimistic than those of the United States), but at any rate it would be reasonable to expect that remaining offensive-defensive arms race interactions would be less dynamic and less destabilizing both in a strategic and a political sense than in the unrestrained case.

Even if such a relative stabilization could be effected, it would have been the result of an offensive-defensive bargaining process which would have involved a great measure of wasted resources in the form of mutually neutralizing ABMs and offensive investments. Expected casualty levels would have decreased on both sides, but – depending on the dynamics of the arms race – this reduction would have been excessively expensive. The political costs in terms of increased international tension, reduced possibilities of crisis management etc., would be a still greater problem.

This fact points to a different model for arms race interactions in the hypothesized setting, namely that of an essentially defensive-defensive model. If the two powers anticipate that a likely result of an offensive-defensive arms race would be some sort of balance around the point where the cost-exchange ratio equals unity (and where their respective assured destruction capabilities would have been reduced compared to the present situation), it would be rational for both powers to concentrate on ABM investments until that point were reached and to avoid neutralizing offensive investments. By doing so, they would also avoid many of the destabilizing effects on the general international situation which would be associated with a vigorous and unrestrained offensive-defensive interaction. The strategic interests of the respective powers would still conflict, but would not be mutually exclusive to the same degree as was the case in the two-dimensional offensive-defensive model.

While such a course would be rational under the prevailing circumstances of the time, it is more doubtful whether it would be realistic to expect it to prevail against the many stimuli that might motivate the two powers to pursue policies aiming at strategic superiority and the like. The defensive-defensive model would in itself constitute a tacit coordination of the strategic arms race that would go beyond anything experienced as yet. To have any validity, it would probably, as a first condition, presuppose a narrowing of the present divergence between the strategic outlooks of the United States and the Soviet Union.

A main argument against the model (and the 'stabilization' model as well) would probably be that they are unrealistic in so far as they do not take enough cognizance of the two powers' requirements for a reliable deterrent. In the hypothesized situation the requirement of the United States for an assured destruction capability of 20–25 per cent of the Soviet population would be difficult to fulfil; in any case the margin of safety would be small, and this would certainly make decision-makers in the United States hesitate at the prospect.

The problems of continued reliable deterrence might, however, be apparent rather than real. It is questionable whether it is reasonable to consider deterrence as such as a function of some fixed, absolute assured destruction capability (as McNamara appears to do) without taking into consideration the magnitude of the threat to be deterred.[13]

One consequence of such a view would be to say – as McNamara has come close to – that some fixed number of deliverable warheads (e.g. 400 one-megaton bombs) would be enough for all deterrence purposes irrespective of whether the power to be deterred were able to threaten with a force of 1000, 100 or 10 nuclear warheads. This proposition seems both historically and intellectually untenable. As a matter of fact, McNamara's arguments about United States deterrence requirements do not seem quite consistent on this point. As noted by D. G. Brennan, 'Mr McNamara appears to believe that he could tolerate up to 80 per cent attrition (of the American assured-destruction forces) and still have an adequate deterrent *against an undiminished Soviet threat*'.[14] Of course, he does not really believe this. If he did take this view, he would have difficulty – especially considering the background of his cost-effectiveness doctrines – to explain why he procured an over-capacity of about 1000 per cent over what was really needed for deterrence purposes. The explanation which he gave in his San Francisco address (the speech in which he announced the decision to deploy the Sentinel system) was much more 'down-to-earth' and reasonable than might have been expected on the basis of his deterrence theory, namely that the present very large strategic forces of the United States had been procured on the basis of calculations (which proved to be over-pessimistic, but that is not the point) of the future Soviet capability to threaten the United States,[15] i.e. the American deterrence requirements had been conceived in relative rather than absolute terms.

A logical consequence of the view that a reliable deterrence is as much a function of some resonable relationship between threat and counter-threat as a function of some fixed damage capability, is that some of the deterrence problems connected with massive ABM deployment would appear less urgent. Deterrence might be reliable even at lower levels than those discussed here, and in theory (but probably not in practice) both powers might allow their damage capabilities to be eroded to very low levels provided they could make sure that the reduction would be even on both sides.

The 'conservative' power (i.e. the one who first became nervous for its continued deterrence capability) would then be in a relatively favourable position to stop the 'chain reaction' and freeze the strategic balance. The same would apply to interactions on the level where the cost-exchange ratio is about 1 : 1, but not necessarily on higher levels of mutual damage capabilities. Thus, the problems connected with uncertainty concerning the strategic balance need not be as large as might be expected.

Fundamentally, a situation in which missile defences and offences compete freely would underscore the interdependence of the strategic postures of the two powers still more than is the case under the present strategic balance.[16] Both powers would have the choice between neutralizing each other's moves in an offensive-defensive interaction pattern or coordinating their moves in a defensive-defensive interaction. From an economic point of view and still more from the point of view of international stability the second pattern would be preferable to the first, because the conflict of interest between the parties would probably be felt as less acute.

The crucial problem is how to arrive at this pattern and to avoid the reciprocal search for security degenerating into a more vigorous and destabilizing variety of the arms race. This is where arms control considerations enter the picture. Arms control problems in an ABM world have not attracted much interest so far, no doubt because most thinking on these problems has been concentrated on the more urgent problem of how to *prevent* deployment of missile defences. We cannot know whether these efforts will succeed, but the real possibility that they may fail should be enough to motivate speculations about how to control devlopments which we may not be able to stop altogether.

In the above discussion it was indicated that a coordination of the Soviet and American weapons programmes might be established in two ways: either on the basis of experience drawn from a vigorous and destabilizing arms race, or by anticipation of the likely outcome of an offensive-defensive arms race. In practice, the latter would presuppose that each power invested its military resources with a view to avoid neutralizing countermeasures in the other's weapons programme. Furthermore, the theory of some sort of stabilization around a balancing point presupposes that the two powers can agree on a level of mutual damage capabilities on which to base the strategic balance. The exact level agreed upon is not important *per se*; what is important is that the powers reach some general agreement or understanding. In this bargaining process the point where the cost-exchange ratio is about unity may be seen as a convenient focal point for the expectations of the parties.

The above-mentioned considerations point to the need for a narrowing of the present gap between the American and Soviet strategic doctrines. For instance, the Soviet Union would have to give up any ideas that they may be entertaining of deploying ABMs as a solution to the problem of strategic dependence on the United States, while the United States would have to give up the idea that the present level of deterrence cannot be decreased.

The first step in an arms control strategy would then be a continuation and intensification of the strategic dialogue both on formal and informal levels. The coming negotiations between the United States and the Soviet Union could – in the absence of more concrete agreements – at least be useful by narrowing the gap or – in the worst case – defining the area of disagreement.

The next step – still assuming that no ban on ABMs has been agreed upon – might be to look for arms control measures that would aim at preventing at least some of the destabilizing effect of ABM deployment. One important obstacle to a formal overall regulation of the strategic arms race is probably the unwillingness of the two powers to tie their hands in a field considered of prime importance to their national security. This would not necessarily mean that more limited measures, which would minimize uncertainties, would be excluded. D. G. Brennan has pointed to the possibilities of informal understandings about overall strategic-force sizes, monitored by unilateral estimates of strategic budgets.[17] Such an understanding would serve to make the two powers more cost-minded and would probably make them less prone to enter into such open-ended commitments as massive ABM deployments would be. Furthermore, it might serve as a brake upon the tendency for ABMs and offsetting offensive investments simply to be added to existing strategic budgets. Another possibility might be to look for tacit or formal understandings, specifying which are 'permissible' reactions to ABMs and which are not. Increased reliance on bombers or cruise missiles have been mentioned as possible ways of evading an opponent's missile defences; these weapons will probably not get any dominant place in the search for continued penetration capabilities (which almost certainly will concentrate on penetration aids and MIRVs), but might nevertheless be regarded as a factor of uncertainty by both powers. These considerations might indicate a possibility of an understanding not to pursue these techniques; an understanding which would be both 'cheap' and – at least marginally – stabilizing. Its greatest value would, however, reside in the fact that it might be viewed as a symbol of the will of the superpower to control the strategic environment and as a possible 'take-off' platform for more far-reaching arms control measures.

The possibilities of such more far-reaching measures should also be explored. J. B. Wiesner, who cannot be suspected of being an ABM adherent, has pointed out that missile defences may render a total missile disarmament more feasible than it is now by reducing the problem of cheating in connection with a disarmament treaty.[18] This proposal seems to be too radical to be realistic. However, the idea that the introduction of ABMs might facilitate a *reduction* of offensive force levels seems worth exploring. If the two powers could 'agree' on reductions of their respective assured destruction capabilities via ABMs, then missile reductions might be seen as a measure which would facilitate progress towards a strategic balance which rests on a lower level of damage capability.

The interest of this chapter has mostly been theoretical and much of the argument has referred to hypothetical situations whose relation to future realities may be questionable. As the political and strategic contours of the future cannot be defined with any certainty, the various models presented cannot have any absolute validity. However, it is suggested that this approach has at any rate a heuristic value in pointing to alternative settings of a continued arms race (some of which are preferable to others) and to methods of arriving at better alternatives.

*Implications for the Arms Race*

On practically every account an overall regulation of the strategic environment which excludes ABMs would be preferable to the models presented in this paper. Prudence, however, dictates that we must take into consideration the possibility that such a regulation will not be agreed upon and that missile defences may constitute integrated parts of the Soviet and American strategic postures in a not too distant future. Such a situation would present a lot of new and unfamiliar problems in connection with the maintenance of a stable political and strategic environment. The sooner we start to think of solutions to these problems, the better will be the chances of avoiding the least desirable of future worlds.

*References*


1 J. B. Wiesner. 'The Cold War Is Dead, But the Arms Race Rumbles On', *Bulletin of the Atomic Scientists*, June 1967, p. 6.

2 Cited in *Survival*, Apr. 1967, p. 114.

3 Cf. N. Talenski. 'Antimissile Systems and Disarmament', *Bulletin of the Atomic Scientists*, Feb. 1965, pp. 26–29; statements by Premier Kosygin in London, 9 Feb. 1967 and in New York, 25 June 1967 (extracts in *Europa-Archiv* 1967/6, pp. D137 ff. and 1967/15, pp. 351 f.).

4 R. L. Rothstein. 'The ABM, Proliferation and International Stability', *Foreign Affairs*, Apr. 1968, p. 498.

5 *The New York Times*, 17 Jan. 1968.

6 J. J. Holst ('Antirobotvapen – hot eller löfte?', *Strategisk Bulletin*, May 1967, p. 26).

7 *Survival*, Apr. 1967, p. 114.

8 R. B. Foster. 'The Impact of Ballistic Missile Defense on Arms Control Prospects', in J. E. Dougherty and J. F. Lehman, eds.: *Arms Control for the Late Sixties* (Princeton, N.J., 1967), p. 88.

9 According to McNamara's *1968 Posture Statement* the Sentinel system will decrease American casualties from 120 to 100 million in case of a Soviet first strike and from 120 to 90 million in case of an American first strike – under the assumption of no Soviet response. Cf. *Statement by Secretary of Defense Robert S. McNamara on the Fiscal Year 1969–73 Defense Program and the 1969 Defense Budget* (Wash., 1968), p. 64.

10 Cf. statements in *Krasnaja Zcezda*, 12 Nov. 1967 and by General Krylov. Quoted in *International Herald Tribune*, 20 Nov. 1967.

11 Hearings on Military Posture and H.R. 9240. House Armed Services Committee, 90th Congress, 1st Session, p. 430.

12 *1968 Posture Statement*, p. 50.

13 The following arguments owe much to D. G. Brennan's paper in D. G. Brennan and J. J. Holst, *Ballistic Missile Defence: Two Views* (Adelphi Paper No. 43, Nov. 1967).

14 *Ibid.*, p. 14. Italics in original.

15 McNamara Address to UPI Editors and Publishers (18 Sept. 1967), *Bulletin USA*, issued by USIS, Copenhagen, p. 5.

16 N. Talenski has pointed out that 'the creation of an effective antimissile system enables the state to make its defenses dependent chiefly on its own possibilities, and not only on mutual deterrence, that is, the goodwill of the other side' (*op. cit.*, p. 28). However, as long as missile defences are not as effective as Talenski presupposes, the opposite seems to be nearer the truth.

17 D. G. Brennan and J. J. Holst, *op. cit.*, p. 16.

18 J. B. Wiesner. 'Hope for GCD?', *Bulletin of the Atomic Scientists*, Jan. 1968, p. 13.

M. LIETENBERG

# 2: The Interaction of Ballistic Missile Defence, the Weapons Testing Programme, and a Comprehensive Nuclear Test Ban Treaty

This chapter assumes some understanding by the reader of the requirements for warhead design and weapons effects testing that would be desirable, or mandatory, in conjunction with procurement of ABM systems. Information related to such requirements has appeared elsewhere and is mentioned in Part 1. The interactions of such a warhead testing programme and the pressures for it, and the decisions that occurred or were required in the United States for or against agreement on a Comprehensive Test Ban Treaty (CTBT) or a Partial Test Ban Treaty (PTBT) will be discussed in this chapter. Most of the discussion concerns the interaction of United States congressional opinion and appearances before that body of various United States administration spokesmen.

*Underground Nuclear Testing subsequent to the signing of the Partial Test Ban Treaty*

Tables 8, 9 and 10 present the essential information. The discrepancy in publicly released figures from official United States and Swedish sources should be noted. Even before 196 and the PTBT, the United States Atomic Energy Commission (USAEC) had stated that it does not announce all United States, or all Soviet tests that it knows of. The AEC has repeated such statements more recently. It has become evident, in addition, that at least for 1964 there are many additional very low yield United States tests whose total does not appear in the table. For example, between 16 January 1964 and 14 Apr

1964, the United States AEC announced five tests. A search in the *Bulletin of the International Seismological Centre*, Edinburgh, Scotland, indicates that eight more tests took place in the same period of time. One can say as an aside that Table 1 bears in some graphic way an implication and insight into the past United States negotiation position on on-site inspections.

The large increase in underground testing has been directly related to anti-ballistic missile warhead development and will be given further and continuing impetus by the procurement of ABM systems. Dr John Foster, testifying in 1966 before the United States House Armed Services Committee on the Nike-X anti-ballistic missile system, indicated that the need for testing would continue: 'Following the deployment of the thin defense one can contemplate building up the defense to be more effective against sophisticated attacks. . . . There is no question in my mind but that a series of experiments involving nuclear explosions would be of great benefit.'

TABLE 8

*Post Partial Test Ban Treaty Nuclear Test Explosions*

| | *U.S.A.* | | *U.K.* | | *U.S.S.R.* | | *France* | | *China* | |
|---|---|---|---|---|---|---|---|---|---|---|
| | *U.S.* (AEC) | *Sweden* (FOA) | *U.S.* (AEC) | *Sweden* (FOA) | *U.S.* (AEC) | *Sweden* (FOA) | *U.S.* (AEC) | *Sweden* (FOA) | *U.S.* (AEC) | *Sweden* (FOA) |
| 5/8–12/31, 1963 | | 14 | 0 | 0 | | 0 | | 0 | 0 | 0 |
| 1964 | 28 | 28 | 1 | 1 | 3 | 6 | ? + | 0 | 1 | 1 |
| 1965 | 28 | 27 | 1 | 1 | 4 | 9 | ? + | 2 | 1 | 1 |
| 1966 | 39 | 40 | 0 | 0 | 7 | 12 | 5 | 5 | 3 | 3 |
| 1967 | 27 | 29 | 0 | 0 | 4 | 13 | 3 | 3 | 2 | 2 |
| *Totals* (1964–7) | 122 | 124 | 2 | 2 | 18 | 40 | 8 + | 10 | 7 | 7 |

Senator Jackson similarly stated, in a speech on the floor of the United States Senate, 30 November 1967 (Senator Jackson is Chairman of the Nuclear Safeguards Subcommittee of the Senate Committee on Armed Services and Chairman of the

TABLE 9

*Total Nuclear Test Explosions* 1945–1.1.68 (*approximate*)

| | *Air* | *Underwater* | *Underground* | *Totals* |
|---|---|---|---|---|
| U.S.A. | 193 | 5 | 245 | 443(+) |
| U.S.S.R. | 161 | 1 | 42 | 204(+) |
| U.K. | 21 | 0 | 4 | 25 |
| France | 9 | 0 | 9 | 18(+) |
| China | 7 | 0 | 0 | 7 |

TABLE 10

*Total Nuclear Test Explosions, 1945–1968*

| | *1945–1963* | *1963–1.1.68* | *Approx. Totals, 1945–1968* |
|---|---|---|---|
| U.S.A. | 319 | 124(+) | 443(+) |
| U.S.S.R. | 164 | 40(+) | 204(+) |
| U.K. | 23 | 2 | 24 |
| France | 10(+) | 7 | 18 |
| China | 0 | 7 | 7 |

Military Applications Subcommittee of the Joint Committee on Atomic Energy):

During the past year the Department of Defense, charged with the responsibility of determining the effects of nuclear weapons, has continued to develop methods of conducting underground tests in which results are being obtained that were previously thought impossible under the treaty restrictions. The accelerated underground test program of the Department of Defense for the next eighteen to twenty-four months consists of a relatively large number of tests on new

> re-entry vehicles, guidance systems, and our anti-ballistic missile systems now under development. As a result the actual detailed test program has developed into a fast moving and changing program because of numerous scientific discoveries and proposals for new testing techniques that are being developed . . . a large number of underground tests were conducted and very significant advances made in the area of weapons technology development, new and radically different weapon design concepts, and in the science of peaceful uses for nuclear explosives. . . .
>
> The basic aims of upcoming underground tests are for the furthering of our knowledge of weapon effects, for improving weapon reliability, increasing penetration capability, and advancing technology.

It was made clear before the United States Senate in 1963 that the United States had a large lead on the Soviets in underground testing experience, technique and capability. It was hoped that with further experience, tests of weapons up to 'several hundred kiloton' might be done underground. Drs Brown and Bradbury nevertheless estimated 'up to one megaton' in 1963. Despite the requirement of emplacing weapon and instrumentation tunnels 3–4000 feet under the ground for these tests, human ingenuity, the intense research effort, and technical instrumentation being what they are, it has turned out to be much easier than anticipated to do larger underground weapons tests, now up to one megaton; and possible to gain much more information than one expected from such underground tests.

Both of these advances are of course exactly what the test ban had been meant to inhibit if not prevent. Senator Jackson has also indicated that several of the Soviet underground tests have been of higher yield than those of the United States. Three of these have vented. Apparently the Soviet Union has not taken the precaution of drilling its shafts deep enough.

Senator Jackson also stated that $30 million had been added to the previous budget for underground testing in fiscal year 1967, and that the Department of Defense has increased their planned expenditures for underground testing by 50 per cent over what they had requested for 1967. About $245 million will be spent for underground testing by the United States in fiscal year 1968 and $314·6 million is allotted for fiscal year 1969.

In order to carry out these new and larger tests, the United States AEC has had to open two new test sites, as the magnitude of the blasts made the old site too close to populated centres.

> The AEC and the Department of Defense determined in mid-1966 that it was essential to establish a capability for conducting higher yield tests underground than was determined to be possible at the National Nuclear Test Site in Nevada. Originally, the Pahute Mesa, at the north end of the regular test site, was thought to be suitable for higher yield tests, but experience disproved this hope and other sites have been selected. The first, still in Nevada, is about 70 miles northwest of Tonopah, Nevada, in an area named Hot Creek Valley. This area is thought suitable for going beyond the yields possible at the Pahute Mesa site. Next, an uninhabited island near the western end of the Aleutian Chain, Amchitka Island is being developed for possible higher yield explosions.[17]

One of the recent tests opened a 4–5000 foot surface fissure in the Nevada desert, but the AEC has indicated it is interested in testing still higher yields and some sources have indicated that a 3-megaton underground detonation would be within safety limits.

As the decision to procure ABMs reached a head in the United States, there were several expressions of fear that it might be considered necessary to abrogate the treaty in order to test certain aspects of the missile system in the atmosphere. For the moment, this fear is expressed as being somewhat less. In the 1967 hearings,[18] Representative Hosmer asked Dr Foster (Director of Defense Research and Engineering, Department of Defense) about a statement he made four years before in testimony:

> Suppose the USSR were to develop a defense such that our ability to penetrate might depend on a saturation attack. For this application specially designed hardened warheads might be required. Considerable progress on such warheads can be made by underground test, but under the treaty, atmospheric experiments to determine if the warhead

actually has the necessary hardness against combined radiation and shock effect would be prohibited. We might thereby be denied assurance of such a penetration capability.

Without saying why, can you be any more reassuring about the situation today than you were in 1962 or 1963, whenever it was?

*Dr Foster:* 1963. Yes, Mr Hosmer. I did indeed indicate that the important sets of experiments that one would not be allowed to do without atmospheric tests was the actual testing of the re-entry vehicle to nearby nuclear explosions in the atmosphere.

Subsequently, we have made very aggressive efforts in our underground program, and I believe that in executive session I can convince you that this is no longer a problem.

Similarly, Dr Foster answered to a question in another hearing:[19]

4. If we decide to build and deploy an ABM system, will it be necessary to resume atmospheric testing?

The answer is no. Our underground test program will assure us of satisfactory ABM warhead design, particularly for light defense deployments. There would be somewhat more uncertainty for heavy defenses (p 23).

And in the same hearing:

*Senator Kuchel:* Now, I listened to your answer just a moment ago, General, which had to do with the Nuclear Test Ban Treaty. I thought the morning newspaper attributed to you a statement that we would have to breach that treaty if an ABM system were to be the policy of our Government.

*General Wheeler:* On the contrary Senator, I believe they confused my statement with the statement made [deleted] by someone else.

*Senator Kuchel:* Well, I am going to tell you we need to straighten that out, because that shook me up.

*General Wheeler:* As a matter of fact, this question of whether we could develop the ABM within the limits of the Limited Nuclear Test Ban Treaty, was asked on Meet the Press last Sunday. I answered 'yes', [deleted].

*Senator Gore*: Well, it shook me up as it did Senator Kuchel, and I promptly sent for a copy of General Wheeler's remarks, and he did not make that statement.

*Senator Kuchel:* So your testimony would be, sir, regardless of the thinness or the thickness of an ABM system in this country, that it could be manufactured and deployed without violating the Test Ban Treaty.

*General Wheeler:* This is correct, sir, and I assure you, Senator, I am not advocating that the United States violate a treaty for this purpose. I do not think it is necessary.

*Senator Kuchel:* Oh, no, I had assumed that you would have suggested it would be in our best interests to notify the other signatories.

*General Wheeler:* That is right (p. 97).

This is confusing as Senator Kuchel seems to be implying by his phrasing, and General Wheeler agreeing, that in the case under discussion the United States should withdraw from the treaty, viz. 'notify the other signatories', rather than violate the treaty.

These remarks, particularly Dr Foster's, imply that it will be possible for the United States to proof-test underground for the ABM weapons effects information that it feels it wants to obtain. This removes some fears for the integrity of the present PTBT. On the other hand, with the procurement of the Sentinel system, the implication is not hopeful for those anxious about the possibilities for a CTBT. To press for such a comprehensive treaty while ABM systems were maintained would imply that nations were willing to freeze a confidence level on ABM destruction effects and ICBM warhead structural countermeasures. In the abstract this is not impossible, but in the real world it does not seem very likely. From publicly available sources it appears that the proposed American Sentinel system is far more extensive than the Soviet ABM system already deployed. The Galosh system around Moscow has not been expanded or extended to other cities, and the Tallinn system has now been excluded from ABM capability. The burden of interest in maintaining underground testing, with past numbers of tests as a criterion, will then appear to remain, as it has been in the past, with the United States.

The uncertainties in the ABM system are not mostly those dealing with weapons effects. The two major uncertainties are: a computer programme capable of dealing with saturation a computer programme written to take account of the actual signatures of re-entry vehicles and dummies (penetration aids and decoys).

Probably, if at all, weapons effects problems can be resolved as well undergound as above ground. If we call this class of uncertainties those that could be resolved by underground tests, the two above are much larger than those which would remain, i.e. those which cannot be resolved by underground tests, and which could be investigated more satisfactorily and cheaper in atmospheric testing. If a missile freeze were signed, the pro-testing position would lose its main arguments, though not all of its arguments unless the freeze were a qualitative one as well. With ABM installed one must remember that there will be pressures for testing the ICBM warheads which must penetrate as well as the ABM warhead. This is well understood by Congress.

> . . . both with respect to ascertaining as much as possible what improvements in ballistic defense the Soviets may be able to make, and what hardening and improvement of our own offensive capability is necessary.
>
> . . . . . .
>
> *Dr Foster:* Mr. Chairman, that is an extremely critical question. As you indicate, we do plan to direct against Kwajalein Minuteman and Poseidon missiles configured so as to represent the most effective means of penetrating ballistic missile defenses, and we will be able to see on the radar, with several radars, just how that attack looks.
>
> We can, at the same time, direct one or more Spartans and Sprint, into the general area and simulate an attack, the intercept of an attack at several altitudes. [Deleted.]
>
> In the last three years we have had a whole series of very complicated experiments which prove, first, that our offensive warheads will work and, second, that they are as hard as we say or if we find them to be vulnerable we fix them, and then measure to see that they are indeed repaired.
>
> So this underground program is a very vital part of maintaining the effectiveness of our offensive force to provide a sure destruction. [Deleted.][20]

Perhaps the way to interpret Dr Foster's earlier remarks is that fears for the integrity of the PTBT would grow serious if any United States administration really wanted to build a large ABM system. One can assume that pressures for atmospheric testing are much reduced now from their levels in 1963, due to the success of underground testing, but one should also probably assume that such pressures are felt differently on the part of those supporting the Sentinel light ABM system and those supporting a heavy anti-Soviet ABM system. From the point of view of some observers it will be very difficult to get the last step, a comprehensive test ban treaty, because one is in effect asking for a cessation of research and development.

*The United States Senate Partial Test Ban Treaty Safeguards*

The opening two paragraphs of the Treaty Banning Nuclear Weapon Tests in the Atmosphere, in Outer Space and Under Water read as follows:

> Proclaiming as their principal aim the speediest possible achievement of an agreement on general and complete disarmament under strict international control in accordance with the objectives of the United Nations which would put an end to the armaments race and eliminate the incentive to the production and testing of all kinds of weapons, including nuclear weapons.
>
> Seeking to achieve the discontinuance of all test explosions of nuclear weapons for all time, determined to continue negotiations to this end, and desiring to put an end to the contamination of man's environment by radioactive substances.

An additional document, though not formally voted on by the United States Senate nor formally appended to the treaty are the 'Safeguards Proposed by the Joint Chiefs of Staff with regard to the Limited Nuclear Test Ban Treaty.'

> A. The conduct of comprehensive, aggressive, and continuing underground nuclear test programs designed to add to our knowledge and improve our weapons in all areas of significance to our military posture for the future.
>
> B. The maintenance of modern nuclear laboratory facilities and programs in theoretical and exploratory nuclear technology which will attract, retain, and insure the continued application of our human scientific resources to these programs on which progress in nuclear technology depends.

C. The maintenance of the facilities and resources necessary to institute promptly nuclear tests in the atmosphere should they be deemed essential to our national security or should the treaty of any of its terms be abrogated by the Soviet Union.

D. The improvement of our capability, within feasible and practical limits, to monitor the terms of the treaty, to detect violations, and to maintain our knowledge of Sino-Soviet nuclear activity, capabilities and achievements.[7]

Criteria to ensure the fulfilment of the safeguards were also spelled out by the Joint Chiefs of Staff.

Senator Jackson provides us with a background to the 'Test Ban Treaty Safeguards':

By way of a quick review, it will be recalled that in 1963, when the Senate committees were reviewing the then proposed Limited Nuclear Test Ban Treaty, the Preparedness Investigating Subcommittee shared with the Joint Chiefs of Staff a serious concern about the Treaty and whether it would serve the best interests of the United States. The Joint Chiefs informed the Senate that in their opinion certain 'safeguards' would be necessary if the Treaty was not to operate against our national security interests. At the request of the Preparedness Subcommittee and the Committee on Armed Services, the Joint Chiefs developed a statement of the specific requirements to implement the necessary safeguards they had defined. . . .

It is significant that the assurances to the Senate given by President Kennedy in August of 1963 that he would fully and effectively implement the safeguards were reaffirmed in their entirety by President Johnson in April 1964.

The Preparedness Subcommittee, because of its role in the formulation of the safeguards, has assumed the role of monitoring the implementation and of making an annual report to the Senate on the implementation. The Joint Committee on Atomic Energy likewise has a deep interest in the safeguards implementation and for the past three years the safeguards monitoring and reviewing has been a joint undertaking. The staff members of both committees follow the safeguards throughout the year and the committee members then conduct a periodic review of progress, the latest of which has just been completed, and this fourth annual report to the Senate on the implementation of the safeguards is a result of that review....

The implementation of the Nuclear Treaty safeguards is the joint responsibility of the Secretary of Defense and the Chairman of the Atomic Energy Commission. To facilitate coordination of the activities of the two agencies in support of the safeguards, the Secretary and the Chairman, in June 1964, formally established joint procedures for the development and periodic review of a National Nuclear Test Program. That program has been developed and submitted to the President, and as directed by the President, plans for implementation are being maintained. The latest White House approval of the current nuclear test program was made in early July 1967.[12]

The author has not looked into the first appearance of the concept or proposal of safeguards, which act as the theoretical underpinning of the continued testing programme. The Preparedness Subcommittee hearings referred to by Senator Jackson took place in May, June and August of 1963. Secretary Rusk, in appearing before the Senate Foreign Relations Committee on 11 March 1963,[6] when negotiations were still essentially concerned with the CTBT, repeatedly referred to a 'safeguarded treaty'. Whether he was referring to Safeguards in the sense of the Joint Chiefs of Staff or referring to on site inspection is not clear; but because the negotiations then dealt primarily with a comprehensive ban, the latter seems more likely. On 11 September 1963, during the Senate debate on the PTBT, President Kennedy wrote a letter to the Senate leaders, Senators Mansfield and Dirksen, regarding the PTBT in which he gave assurances about American nuclear defences. In this letter he touched on all of the safeguards, as well as discussing other items:

1. Underground nuclear testing, which is permitted under the treaty, will be vigorously and diligently carried forward, and the equipment, facilities, personnel and funds necessary for that purpose will be provided. As the Senate knows, such testing is now going on. While we must all hope that at some future time a more comprehensive treaty may become possible by changes in the policies of other nations, until that time our underground testing program will continue.

2. The United States will maintain a posture of readiness to resume testing in the environments prohibited by the present treaty, and it will take all the necessary steps to safeguard our national security in the event that there should be an abrogation or violation of any treaty provision. In particular, the United States retains the right to resume atmospheric testing forthwith if the Soviet Union should conduct tests in violation of the treaty.

3. Our facilities for the detection of possible violations of this treaty will be expanded and improved as required to increase our assurance against clandestine violation by others.

. . . . . .

7. This Government will maintain strong weapons laboratories in a vigorous program of weapons development, in order to ensure that the United States will continue to have in the future a strength fully adequate for an effective national defense.[22]

It would be profitable to consider the possible origins and pressures for the Safeguards. In Dr Teller's appearance before the Senate Foreign Relations Committee[7] he is the first to levy the charge that the Livermore laboratories have been handicapped in their test programme for political reasons; he intimates that it is the Office of the President that is doing the hindering, and challenges the Senators to ask Dr Foster, then Director of Livermore, if this is not so when he appears before the Committee. Teller also makes clear in his testimony before both committees that he has in the days preceding his appearance been to visit various Senators. Having in a personal visit in 1957 caused President Eisenhower to moderate his enthusiasm for a test ban by promising the advent of 'clean' bombs, and having thrown a minor monkey wrench into the Geneva proceedings in 1959 with the decoupling theory, Dr Teller is anything but a novice in these matters. There then occur very interesting exchanges on this exact issue of political limitations of the test programme with Dr Foster, Dr Brown and Dr Bradbury (Director, Los Alamos).[7] Dr Brown had been the Director of Livermore before becoming Department of Defense Director of Defense Research and Engineering. Under leading questions by Senator Jackson, Dr Foster admits that 'During the last year, the laboratories have been under severe non-technical limitations. . . . We have had technical restrictions' and he introduces Dr Bradbury's name tangentially in his reply. Dr Bradbury had just been questioned as the preceding witness and stated he had had no limitations or restrictions. Dr Brown categorically denies that there were *any* limitations, in a year that saw about ninety-five underground tests, about two per week, and implies that Dr Foster is referring to vague project proposals that Livermore never got down as formal project requests. Dr Brown adds that the National Security Council, not known for its political pressures for accommodation with the Soviet Union, made the decisions on all test programmes; and that, further, the 'Teller proposal' on limitation of atmospheric test to a total of 1 megaton of fission products – which Dr Teller had been offering as an alternative on and off for years in his testimony against a test ban – would have prevented the 1962 American atmospheric test series.

There is also a very neat split between McNamara, Brown and Bradbury on one hand and Teller and Foster on the other hand concerning their assessments of the 1961–2 Soviet test series and who is 'ahead' in tests and knowledge derived from tests, particularly knowledge of applicable to ABM problems. Dr Teller simply insists that he feels sure or is sure that the Soviets are ahead. Only McNamara and Brown offer evidence and criteria for their assessments in terms of whether and how many analogous tests have been carried out by the United States and what may have been learned. In Dr Bradbury's words:

> They have conducted in my understanding, and you can get this information from other sources better than you can from me, some experiments of considerable missile sophistication, but not very many.
>
> Very few I should think, and the primary type of experiments they have used is similar to the classical method of airdrop, observing its method and the yield. They do it on occasions and under circumstances which doesn't seem to make it obvious that they have a very elaborate diagnostic program.[7]

Public evidence on the question of whether the Safeguards were an initiative of the Joint Chiefs, or were suggested to them, is unavailable. There is a widespread notion that the administration assumed that the Safeguards were the Senate's 'price' – or that they actually were the Senate's price—for approving the PTBT. There is little question but that they were the Joint Chiefs of Staff's price. Without the Safeguards, the Joint Chiefs of Staff would have balked, implying that the Treaty endangered American security, and with the Joint Chiefs balking, there would be far more chance of the Senate balking. Though it was seldom said in so many words for public consumption, Senator Jackson, in reading the Safeguards, put it as 'the Chiefs in laying down the conditions that they said were, in effect, conditions precedent on their part to acceptance of the Treaty. . . .' The aforementioned letter from President Kennedy begins as follows:

> I am deeply appreciative of the suggestion . . . that it would be helpful to have a further clarifying statement about the policy of this Administration toward certain aspects of our nuclear weapons defenses under the proposed test ban treaty now before the Senate. I share your view that it is desirable to dispel any fears or concerns in the minds of Senators or of the people of our country on these matters. And while I believe that fully adequate statements have been made on these matters before the various committees of the Senate by the Secretary of State, the Secretary of Defense, the Director of Central Intelligence, the Chairman of the Atomic Energy Commission, and the Joint Chiefs of Staff, nevertheless I am happy to accept your judgment that it would be helpful if I restated what has already been said so that there may be no misapprehension.
>
> In confidence that the Congress will share and support the policies of the Administration in this field, I am happy to give these unqualified and unequivocal assurances to the members of the Senate, to the entire Congress, and to the country.[22]

One further consideration relative to the safeguards requires mention. Hearings in both the Senate Foreign Relations Committee and the Senate Armed Services Committee contain many accusations against the Soviets for breaking of the test moratorium in September 1961. They are made by Senators for and against the PTBT. They are a constant political refrain, and in these discussions came to serve as the most immediate reference guide to Soviet untrustworthiness, substantiating suspicion of Soviet intention to cheat, and hence the motive for the Safeguards. This conception of breaking of the moratorium has also served a second but related purpose. The belief has led to a willingness to assume the probability of Soviet cheating, and thus directly to the position on on-site inspection demands in relation to a test ban treaty. This understanding of the end of the moratorium has been accepted by all United States students of arms control problems but one.

However, Inglis has written:

> The limited test ban was signed in 1963, in an atmosphere of suspicion arising from the fact that an informal moratorium on testing had been abandoned a couple of years before, first by the declaration of the United States and then by the sudden action of the Soviet Union, and as a consequence, there was fear that a formal treaty might be no more lasting;[17]

and elsewhere,

> It is too often forgotten that President Eisenhower very explicitly terminated the temporary agreed moratorium on nuclear testing at the end of 1959 and that Khrushchev followed suit in declaring himself free to resume testing, so that we should have not been as surprised as we were by the Soviet massive resumption of testing almost two years later. The myth that the Soviets thereby broke an agreement is still being exploited by opponents of a test ban.[9]

It would be rather important to have a clearer understanding of the role and meaning of President Eisenhower's statement. One begins to wonder if the Soviet breaking of the moratorium really can be considered a cause of the demand for the Safeguards, or a convenient rationalization for the interests of the Joint Chiefs of Staff in continued testing. In addition the stress on the Soviet 'breaking' may have some psychological meaning as a scapegoat of another lesser sort. Thus Inglis also writes:

> The remarkable aspect of that episode is that our military

> apparatus, one part of which was sqawking loudly about the dangers of elaborately contrived, small underground violations of a test ban treaty, was at the same time, without any test ban agreement to assure Soviet restraint, preparing no more than small and militarily insignificant underground tests. While the ostensible reason was to leave a favourable climate for the negotiations, we were retreating further from realistic negotiating positions. The real reason must have been largely that no profound technical urgency was felt for large-scale atmospheric tests at the rather high level of nuclear development that we had then attained.
>
> Yet it must be recognised that failure to go on testing when there is no test ban agreement is an anomaly. Our indulgence in the luxury of this militarily careless anomaly for the two years 1960–1, with no calamitous effects, demonstrates how terrifically exaggerated are the fears of small-scale underground evasion of a test ban. From the point of view of weapons development it would be far more sensible for the long pull to avoid such unilateral carelessness and at the same time to buy the protection of a test ban agreement with its much smaller risks of development advantage for the other side.[9]

However, by the end of the 1961–2 Soviet test series, the Department of Defense may have been worried and perhaps felt it had been caught short. The Soviets had detonated as many tests as they had from 1949 to 1958, and several times the number of atmospheric tests that they had carried out in the longer-period – essentially because they had done relatively few such tests in the earlier period. However only access to classified material could clarify how much of the American margin in weapons design, if any, may have been eroded by this Soviet test series, or how much the nature of the remarks in 1962–3 were demanded by the domestic political scene. The present PTBT contains a clause permitting a nation to withdraw from the treaty if its national security interests so demand. This is ordinarily assumed to permit interpretation by one side of its position in relation to the other side in strategic weaponry, or in some novel development of weaponry. There has been no discussion of whether the breaking of the moratorium by the Soviets can or can not be construed under such a framework.

If it could, it would at any rate have to be discussed in terms other than trustworthiness or untrustworthiness.

*Abrogation or Cheating under a Test Ban Treaty*

It is true that no amount of reasoning or analysis of pressures and motives would undo the effects of an abrogation of the PTBT if it occurred. At the same time there has been a noticeable absence of sensible analysis concerning why a signatory might be interested in holding to the treaty, or under what conditions he would feel impelled to abrogate or cheat. There have often been intelligent, though brief, remarks on this matter by Department of Defense officials at hearings before congressional committees. However, they have always been coupled with the strongest possible recommendation for the Safeguards programme. There is some degree of logical contradiction between these two positions.

The following is an extensive quotation from testimony in 1963 by AEC Commissioner Seaborg:

> We must always remain alert to the fact that one side may try to acquire a superior advantage through violation or abrogation of the treaty. The effect of such an action on the other parties is decidedly less where underground testing is permitted and where an active program of worldwide nuclear test detection is continued.
>
> It does not seem possible to be forewarned against a surprise abrogation. Even if the 3 months' notice period for any country planning to withdraw is given, it would possibly have been accompanied by an earlier period of preparation.
>
> Consequently, we will pursue the most promising directions of development permissible under the treaty and will maintain a state of readiness for conducting tests in the atmosphere and other media. This readiness posture has been stated as national policy by the President.
>
> To minimize the slowing down of scientific developments which would otherwise provide advantages to a nation considering withdrawal, we will continue vigorous programs of development and underground testing by the weapons laboratories with strong support by the Government through the provision of necessary advanced facilities and equipment and adequate supporting staff.

Under the test ban treaty, the Nevada test site and possibly other continental sites will be maintained on a continuing operational basis in order to support the test program which must be carried out.

The maintenance of a state of complete test readiness, however, cannot be satisfied by the maintenance of only underground test sites.

If another nation should abrogate, we would find it necessary to carry out promptly atmospheric and other tests – first, proof tests of warheads and systems, along with limited effects tests because they can be staged more quickly, then weapons developments tests and the more complex weapons effects tests.

The major difference between maintaining a readiness to conduct proof tests and to conduct development tests lies in the fact that for the development tests the devices themselves must be developed to the state where a test serves the maximum purpose. This is the primary function of the weapons laboratories.

The simplest proof test requires a safe detonation area which may be over either land or water, a vehicle for delivery of the warhead, means for determining yield, a rather large number of technically trained personnel, and logistical support.

Similar requirements exist for carrying out development tests in the atmosphere; however, in addition, more complex diagnostic instruments will usually be required as will a complex communication system both to provide for safety and to facilitate the collection of data. Additional technical personnel and greater logistical support are obviously required.

Thus, the readiness to resume a full-scale weapons development program after a surprise treaty abrogation involves two independent, though related aspects – the development of devices and experiments to be used in tests and the maintenance of a capability to carry out tests.

As I have previously stated, the weapons laboratories will play a major role in maintaining a state of readiness through their development of new devices. Maintaining a mechanical readiness to conduct field tests on relatively short notice is difficult.

Retention of a high technical test capability, however, will be assisted to a considerable extent by the continuation of an

> intensive underground test program. Also, under the determined and vigorous policy which we expect to follow, no nation will be able to gain more than a temporary advantage through surprise abrogation.[7]

An incidental problem which is raised by this posture is that it would be impossible to separate these safeguards from the requirements for preparations to test by a nation intending to abrogate.

There seems to be a crossroad where undergound testing, the Safeguards, and the Plowshare Program (Peaceful Uses of Nuclear Explosives) all meet. This has generally been overlooked but may contribute some understanding to pressures behind Plowshare, as well as indicating a further general desire for underground testing aside from its role in weapons development in general and ABM development in particular.

In Dr Seaborg's Senate testimony he states:

> Our ability to maintain a vigorous, imaginative, and productive program including the development of the necessary scientific and technical know-how is dependent on the retention of able scientists and engineers in our weapons laboratories and the attraction of new ones to the program. They must be challenged by new problems, work in a stimulating environment with adequate equipment, facilities, and other support, and have maximum opportunities consistent with national policy to test the products of their laboratory experiments and calculations.
>
> We will strongly support these laboratories and will encourage the planning and execution of the programs, including allowed testing, necessary to meet our requirements.

And in this summary:

> The weapons development laboratories will be provided the support necessary to maintain strong programs, including test support activities. Attention will be given to the provision of equipment and facilities required for a modern effective research and development undertaking and for maintaining the confidence and morale of the associated scientists and other personnel.

Continuation of Plowshare development in these laboratories will contribute to their viability.

And in response to questioning:

(*Question*): In the days before the voluntary suspension of tests, the days before the efficient testing of nuclear power underground, it was said if we could not test in the atmosphere we would not be able to hold our scientific teams together. Has this situation been ameliorated by reason of our more extensive underground testing?

*Dr Seaborg:* Yes . . . I think that there will be a problem in keeping the laboratories strong and keeping a sufficient number of scientists together but this will be helped under this test ban treaty because of the continuance of underground testing.

The same theme occurs in a reply by McNamara to questioning:

*Senator Kuchel:* President Kennedy, in his speech to the nation of over a year ago, indicated that we would never be caught short again, as you and I hope and pray, and we are assuming that we are taking arrangements we will not be caught short again. Secondly, he indicated that there would not be a stable of American scientific brainpower available if tests could not be seen down toward the end of the road.

Would it be your statement that since underground testing is not prohibited in this treaty that the statement of the President ought not to apply here?

*Secretary McNamara:* Yes, Senator Kuchel. I think it is quite a different situation where we will be able to carry on a large number of development projects through underground tests on the one hand, versus complete prohibition of tests in all environments on the other hand. In the latter case, it would be far more difficult to maintain the vitality of the laboratories than it would be in the former case.[7]

As is evident here and in the Safeguards statement, maintaining the weapons laboratory staffs in optimum shape is a very important goal, and testing will obviously help to further that goal.

This is not the place to discuss the interactions of the Plowshare programme, a comprehensive test ban treaty, and, in addition, the non-proliferation treaty. This has been done elsewhere.[25] Suffice it to say that there are obvious interactions. Private discussions with an ex-AEC commissioner as well as public statements of various AEC officials have made clear the AEC approach to the Plowshare programme. For example, Dr Seaborg in 1965:

> I would recommend that such a treaty make it possible to include Plowshare shots. . . . I would recommend the inclusion of a provision in the comprehensive test ban treaty that would allow Plowshare.
>
> (*Questioner*): I take it from your statement you feel that nothing in this treaty will seriously retard that program.
>
> *Dr Seaborg:* That is right, for a number of years. However, in a few years the treaty would require an amendment, if we were going to be able to go forward at that time to do the things that the technology would permit by that time.[16]

And in his summary:

> Promising applications of nuclear explosives for peaceful purposes will be developed and demonstrated under the treaty. At such times as the benefits have been demonstrated, the parties may wish to modify the treaty to permit extension of the technology and benefits to all.[16]

This view was also contained in one of the points in President Kennedy's letter:

> 8. The United States will diligently pursue its programs for the further development of nuclear explosives for peaceful purposes by underground tests within the terms of the treaty, and as and when such developments make possible constructive uses of atmospheric nuclear explosions for peaceful purposes, the United States will seek international agreement under the treaty to permit such explosions.[22]

It has been continued most recently by Dr Seaborg in the Non-Proliferation Treaty hearings before the Senate Committee on Foreign Relations:

> *Senator Carlson:* Doctor, as you well know, the construction of a new canal or improvement of our present facilities at the Panama Canal is a very current issue and one that concerns not only our own Nation but many other nations.

What chance would we have of getting modification of the Test Ban Treaty restriction? Let's assume we were ready to use some of this material to build a sea level canal across the country. Whom would we confer with, all of the countries involved in this treaty? Whose approval would we have to get?

*Dr Seaborg:* I think in the case of the limited nuclear Test Ban Treaty, an amendment requires an affirmative vote of a majority of all the parties to this treaty including the votes of all of the original parties.

*Senator Carlson:* In other words, we built the Panama Canal, and if it now was decided that we, as a nation, wanted to build it we could not go ahead and build it under this treaty without getting the approval of the majority of the nations in this world?

*Dr Seaborg:* It wouldn't be the Non-proliferation Treaty that would be limiting this; it would be the limited Test Ban Treaty.

*Senator Carlson:* The limited Test Ban Treaty?

*Dr Seaborg:* Yes, the limited Test Ban Treaty would have to be amended in order to build a sea level canal across the Isthmus using nuclear explosives. But I do believe that the atmosphere of interest in such uses, and the increased confidence throughout the world that such a project could be carried out without any menace to the public health and safety is creating the kind of a climate that would make such an amendment to the limited Test Ban Treaty feasible, and I believe that the very act of our engaging in discussions concerning the Non-proliferation Treaty has contributed to that climate. It has become more apparent to more and more nations that we can have the benefits of the peaceful uses of nuclear explosives with essentially no detrimental effects concerning the public health and safety.[22]

Something further should be said about the insistent presumption of Soviet cheating or the likelihood of any nation's cheating for that matter under the terms of an arms control treaty, particularly one on which everyone's gaze is riveted because of its being the first of its kind in the post World War II nuclear era. A thorough treatment of this subject should be done: only a devil's advocate position will be sketched here. It should be constantly kept in mind in this discussion that the crux of the matter is the balance between the level of seismic detection capability and the amount of weapon design related information that can be obtained from test yields low enough to escape detection. Several technical points were raised in the various hearings and some of these are now discussed.

The potential violator of an underground test agreement does not know the other side's seismic detection capability and must go well below what he expects it to be unless he wants to run a substantial risk of being detected. Some few sites have substantially better seismic detection capabilities than the average, by a factor of 2 or 3, raising the risk still further. One such especially good site, Mould Bay in the Canadian Arctic, was recording American tests in Nevada down to one kiloton in alluvium in 1963. Such low yield tests as would be required to be sure to escape present detection capabilities supply relatively little militarily useful information, so little that it is often said that if the test escapes detection one need not worry that the information derived from it will alter the strategic balance between the two powers. In a sense then, one might consider permitting 'undetectable cheating'. Finally, more useful information is usually derived from a series of tests, which would stand more chance of being detected than an individual test.

In 1963 the spectre of Soviet cheating in the upper atmosphere was also seriously raised by some witnesses (Dr Teller). In mid-October 1963 an Atlas Agena rocket successfully placed into orbit two instrumental satellites designed for the detection of nuclear explosions in space. Additional satellites with the same function have been placed into orbit subsequently. In addition, work continued on ground based detectors of nuclear explosions in space. This is therefore now guarded against by ground and by satellite detection methods, both for vehicle launch and ex-atmospheric detonation; but it was also pointed out in 1963 that to avoid detection and gather useful data would be very difficult. The costs would be enormously extravagant, and for the cost the information gathered would be minimal.

Aside from these technical considerations which were covered in some detail during the hearings by both McNamara and the

Joint Chiefs of Staff, there are at least equal, if not vastly more important political considerations. It has been stated that an important pressure behind the interest of many nations in a test ban treaty was to reduce contamination from radioactive test debris. Concepts of cheating, and the fears of its possible strategic consequences are irrelevant if the main motive for the negotiating powers was the limitation of environmental pollution. If one could not assume that fears regarding cheating are related to weapons aspects, as they are obviously not related to pollution, one is entitled to immediately disregard any arguments presented on behalf of such fears. These could only be taken seriously in regard to weapons aspects, and therefore in order to discuss them, the following discussion presumes the motive of arms control for the PTBT.

A major purpose of the treaty to most United Nations members was to slow down the nuclear weapons race. Should one side abrogate or cheat and be detected, the other side would resume testing as well, and this purpose would be lost.

The second major purpose of the treaty to most United Nations members was precisely to develop trust between the major powers so that an environment could be developed in which further arms control agreements could be entered into. Again the nation that abrogates and/or is caught cheating would eradicate this purpose. It does not seem sensible to suppose that a nation would negotiate in an international spotlight for 4–5 years intending all the while to torpedo the outcome subsequently. Since the entire negotiations were carried on in the Eighteen Nations Disarmament Conference under United Nations auspices, prodded by resolutions voted on by the entire United Nations body, that nation that abrogated would pay severely on the international diplomatic front. If all it wanted to do was generate good will among other United Nations powers during the negotiations, then it definitely should not sign the treaty in the end. Breaking a treaty subsequently would engender more ill will in the opposing powers than simply not ever having suggested or taken up the matter in the first place, and simply going on testing.

It is obvious that this very price is the best reason that can be offered for cheating, as opposed to abrogation. But if a nation decides to cheat it is deciding to run the risk of being detected, which would probably earn it greater international condemnation than simply abrogating. Therefore, in deciding to cheat, i.e. test, it might as well abrogate and do it openly, earn less condemnation, test at a higher yield, at least gather more useful data, and do it much less expensively in financial cost. From a pragmatic standpoint then, in Dr Bradbury's words,

> To me it simply seems the hard way to do something which, if you are determined to do it, you might just as well abrogate the treaty and go ahead and do it. You will probably be found out doing it by one detection system or another, or even by intelligence methods. Is it worth taking that risk of being caught up when, as I say to be honest is cheap and only requires 90 days.[7]

If one major nuclear power broke the treaty it would be harder than ever to get the other to agree to anything again. The military consequences would be most serious for the weaker of the two powers, and in most estimates that is the Soviet Union.

Little mention is ever made of the fact that the Soviet Union is interested in limiting further United States weapons development. Anything the Soviet Union can do, the United States can do at least equally well. Yet the Soviet Union seems never to raise the matter of American cheating, nor does anyone in the United States. Aside from the technical risk of being caught, the enormous official American and Senatorial stress on cheating seems at best enormously overdrawn, at worst entirely meaningless.

### *The Present, Future, and Other Related Matters*

It is often said that the political motivation for a three environment test ban was to limit radioactive pollution. This belief exists on the part of many, and it can only be stated, and noted that it is in some contradiction to the discussion of the treaty in this chapter solely in terms of its weaponry aspects.

Of all arms control measures, perhaps the main international support of the smaller and non-nuclear nations is for a comprehensive test ban; they have become educated on the technical background of the issue perhaps more than on any other. It would directly respond to their request that the nuclear nations restrain their weapons programme.

Whether the United States government at any time intended to sign a comprehensive test ban including limitations on underground weapons tests seems to be an open question. The historical evidence is reviewed elsewhere.[25, 1, 2, 3] The 1963 'mistake' in Geneva on the number of inspections[3, 22] might have occurred and cannot be absolutely ruled out. It is very unlikely, but not impossible. There are very strong indications that the American test detection capabilities are better, or far better, than have been publicly admitted. The classified American test detection systems add an additional unknown quantity to this capability, and it is not too difficult to guess at what the classified capability might comprise. McNamara's testimony, carefully analytical in terms of its military advantages to the United States perhaps needs some evaluation in terms of the nature of its audience and purpose. Research and development remain the very last things to be limited in disarmament postulates. Military establishments will be very reluctant to give these up, and that is what underground testing is. Such reluctance will appear in the split in motives, in utilities to be expected and anticipated.

There is surely some evidence which speaks against such American intention. Little of the evidence derived from testimony by official United States government spokesmen in support of the partial treaty can be gratuitously assumed to simultaneously apply to what the same spokesmen might have said in support of an undergound test ban as well. Some officials and departments that supported the partial treaty, with or without reservations (the Safeguards), surely would not have supported an underground ban as well. It is possible that the United States would have signed either treaty, but that the Senate might not have ratified either treaty. In that case would the government have signed a treaty on which there were doubts about its ability to insure ratification. There are no answers to these questions without access to authoritative sources. Without such access one may be left free to assume that the issue was continuously in doubt, that a final decision may not ever have been reached, and that certain major sectors in the administration that were at varying times known to be against an underground ban may have been strong enough to stalemate the situation, if not to have caused a decision in the negative to be held. There is no evidence that would detract from the conclusion that the official United States government position was to sign a three environment (partial) test ban treaty once it was willing to consider such a ban alone and not as part of a package proposal containing other measures.

Nearly all discussions of the test ban treaty imply or state explicitly that development of nuclear weapons is impossible, or nearly impossible, whatever that means, without testing. Discussion of this topic is impeded, not necessarily unhappily, by classified information, yet it has never had adequate treatment in the arms control literature. Kramish states:

> Experience gives us no reason to expect that a nation's first atomic device, based upon a simple and tried design, will fail. If a nation is reasonably confident that it has the recipe for the Nagasaki bomb, and feels confident of the results of preliminary testing of non-nuclear components, we can be reasonably sure that its device will explode. . . . Only by dabbling in advanced nuclear designs, seeking to increase efficiencies, adapting warheads, etc., does one incur substantial risk of failure.

Without training in nuclear physics and experience in the inner councils of the *N*th nations it is impossible to reconcile these very divergent positions, and it seems to me to be a desirable issue on which to have as little uncertainty as possible. People with experience in weapons design often tend to feel that a crude design would work, without testing, but simultaneously feel that no military would be willing to use an untested weapon. To what degree such feelings derive from ways of military thinking in the major powers, and perhaps less resolute elsewhere is unknown. To what degree such opinions will hold 25, 50 and 100 years from now is also unknown. An untested weapon would have less credibility in the context of deterrence, less nuclear club impressiveness, and the opponent could always hope it would not work, but it nevertheless might have some relevance in all these areas. The degree to which all observers agree that nuclear weapons are within the technical capabilities of the first rank *N*th nations is an indication of the evaluation problem that would arise if 1, 3 or 6 of these countries announced after *x* years that they had fabricated an untested weapon. This cursory discussion may not be on the firmest of grounds, but these are problems that are likely to arise with time, and the interest and value in halting proliferation of nuclear weapons is not relegated to the next 10 years only, but to the next 100 as well.

There seems to be a direct connection between Plowshare and the Safeguards programme.

With above-ground testing each side was at least able to monitor the other's atmospheric tests, and from fission products and other data, was able to make some assessment of the other side's weapons development. In addition to 'designing around' resulting uncertainties for offensive weapons systems by redundancies and improvements on individual components, which in effect serves to increase the effectiveness and lethality of offensive systems, inability to monitor the other side's underground tests may also have produced an untoward effect further accelerating underground testing programmes. Uncertainty breeds insecurity in arms races, insecurity breeds fear, and fear over-compensations; one such over-compensation perhaps being excessive underground testing.

Maintaining maximum readiness to test as a safeguard fulfils simultaneously every requirement for maximum readiness to abrogate. The Safeguards programme, albeit designed to obviate being 'caught short' again should the other side break the treaty, as the United States indicates occurred in 1961–2, at the same time seems to be a classical example of how to run the maximum risk on 'how does it look to the other side', i.e. in being read as a signal that was not intended.

*Conclusions*

One is left with the question of exactly what the partial test ban treaty may have achieved of its original intentions as regards the major nuclear powers. As far as inhibiting testing, retarding offensive nuclear weapons development and ABMs in 1968, five years after, the answer is absolutely nothing, unless one assumes that the situation would now be yet worse if atmospheric testing had continued. The treaty may in the future have some effect on making development prohibitive or at least far more difficult for non-nuclear powers, unless such nations renounce the treaty.

There seems to be some analogy possible between the partial test ban treaty and the post World War I arms race situations, in that certain nations were willing to request limitations for others but not for themselves. If the analogy is acceptable, the lesson from 1920–35 is that this is not the most likely avenue to success. It was recognized in 1963 at the signing of the PTBT that unless further progress was made, the benefits from this treaty would diminish with time. As regards the major nuclear powers, the benefits have not only been diminished but totally eradicated, and probably rather faster than anyone had expected. Again in the negotiation of a non-proliferation treaty the major nuclear powers have disregarded the repeated requests of the 'eight' in the Eighteen Nations Disarmament Conference and of other non-nuclear United Nations members for agreements for limitation of offensive missiles, limitation of fissionable materials production, prohibition of ABM, reduction of nuclear weapon stockpiles, etc., in short, on moves which abate or even invert the direction of the United States–Soviet Union nuclear arms race.

Efforts between 1963 and 1966 at Geneva on a CTBT showed some commendable aspects on both sides, but there have been no further discussions since 1966. With the procurement of ABM systems one cannot assume that either the United States or the Soviet Union is at the moment much interested in achieving 'the discontinuance of all test explosions of nuclear weapons for all times'. If a comprehensive test ban treaty were signed it would be a major advance, and a major signal concerning the intention of the nuclear powers for the future of the arms race. If it is not, the outcome of the corollary agreements to the Non-proliferation Treaty may turn out to have the greatest significance to the viability of the PTBT, and to the motives behind its formulation. For the interaction between 'the provisions and the preamble' of the NPT will have much to do with the outcome of its possible review each five years, and its 25-year lifetime, in a way in which the three environment test ban has not had to face being tested. The life of the NPT will in fact be dependent on the success of further arms control measures. In this light ABM and continued underground weapons testing in relation to it are implicitly and explicitly in contradiction to the NPT and bode poorly for the future.

## *References*


1 *Diplomats, Scientists and Politicians*, H. K. Jacobson and E. Stein, 1966.

2 *Postwar Negotiations for Arms Control*, B. G. Bechhoefer, 1961.

3 *The Neutrals and Test Ban Negotiations, An Analysis of the Non Aligned States Efforts between 1962–1963*, M. Samir Ahmed, 1967.

4 *Negotiating Disarmament, ENDC The First two Years 1962–1964*, A. S. Lall, 1964.

5 *American Scientists and Nuclear Weapons Policy*, Robert Gilpin, 1962,

6 *Test Ban Negotiations and Disarmament*, Hearing, Committee on Foreign Relations, U.S. Senate, 11 Mar. 1963.

7 *Nuclear Test Ban Treaty*, Hearings, Committee on Foreign Relations, U.S. Senate, Aug. 1963.

8 *Military Aspects and Implications of Nuclear Test Ban Proposals and Related Matters*; Part I and Part II, Hearings, Preparedness Investigating Subcommittee, Committee on Armed Forces, U.S. Senate. May, June, August 1963. Also *Interim Report on the Military Implications of the Proposed Limited Nuclear Test Ban Treaty* (same committee and subcommittee), Sept. 1963.

9 *Test Ban Default. Obstacles to a Crucial Agreement*, D. R. Inglis, Council for Correspondence, No. 27, June 1963.

10 *Underground-Test Ban Debate*, *Scientific American*, Aug. 1966.

11 Speech, H. H. Humphrey, Senator, U.S. Senate, 7 Mar. 1963.

12 *Remarks on Nuclear Test-Ban Treaty Safeguards.* Henry M. Jackson, Senator, U.S. Senate, 30 Nov. 1967.

13 Letter from Secretary of Defense McNamara and AEC Chairman Seaborg to President Johnson: Implementation of Test-Ban Treaty Safeguards, 16 Apr. 1964, *Documents on Disarmament*, 1964.

*Implications for the Arms Race*

14 *Nuclear Weapons*, Sir John Cockcroft (in) *Unless Peace Comes*, Nigel Calder (ed.), 1968.

15 *The Outlook for Nuclear Explosions*, David Inglis (in) *Unless Peace Comes* (ed.), Nigel Calder, 1968.

16 *Peaceful Applications of Nuclear Explosions, Plowshare*, Hearings, Joint Committee on Atomic Energy, U.S. Congress, 1965.

17 *Peaceful Nuclear Explosions and Disarmament, Plowshare, Proliferation and Testing*, Thomas S. Lough, 30 Aug. 1967 (private distribution).

18 *Prospects and Problems. The Nonmilitary Uses of Nuclear Explosives*, David R. Inglis and Carl L. Sandler, *Bulletin of the Atomic Scientists*, 23, No. 10, Dec. 1967.

19 *Scope, Magnitude and Implications of the United States Antiballistic Missile Program*, Hearings, Subcommittee on Military Applications, Joint Committee on Atomic Energy, 6–7 Nov. 1967.

20 *United States Armament and Disarmament Problems*, Hearings before the Subcommittee on Disarmament of the Committee on Foreign Relations, United States Senate, Ninetieth Congress, 3, 6, 7, 28 Feb. and 1, 2, 3, Mar. 1967.

21 *Non-proliferation Treaty*, Hearings before the Committee on Foreign Relations, United States Senate, Ninetieth Congress, Second Session: Treaty on the Non-proliferation of Nuclear Weapons, 10, 11, 12, 17 July, 1968

22 *Documents on Disarmament 1963*, U.S. Arms Control and Disarmament Agency; Publication No. 24, Oct. 1964. pp. 489–91.

23 *Kärnladdningsexplosioner 1945–1966*, I. Zander and R. Araskog, FOA 4 (Swedish Defence Research Inst.) Report. A 4493–16, June 1967.

24 *A Comprehensive Nuclear Test Ban. Technical Aspects 1957–1967*, Christopher Hohenemser, and Milton Leitenberg, *Scientist & Citizen* **9**, No. 9–10, Nov.–Dec. 1967.

25 *United States Attitude Towards a Comprehensive Nuclear Test Ban Treaty: History, Relevant Factors, and Present Position*, Milton Leitenberg, Third Nordic Conference on Peace Research, 22 May 1968.

D. R. INGLIS

# 3: Region-by-Region Disarmament with Anti-Ballistic Missiles

At a Pugwash Conference in Moscow in 1960 L. B. Sohn presented a proposal for disarmament by regional stages, an idea conceived independently also by A. P. Alexandrov and Melvin Mooney, as a very promising means of avoiding the incompatability of military secrecy and verification by inspection in the process of stage-by-stage disarmament. This idea seemed to remove the main technical difficulty and leave essentially only political difficulties in the way of progress towards disarmament. This plan subsequently found its way into a United States disarmament proposal in an incidental way but seems to have received no serious consideration in recent years.

J. B. Wiesner[1] points out that the deployment of ABM systems, which seems almost inevitable, may have an important bearing on the modalities of disarmament. The low-level stabilized deterrent has been discussed as an important way-station on the road to general and complete disarmament (GCD). Some resistance to it stems from the fear that it may not be able to cope with clandestine preparations. Wiesner points out that even the 'thin' ABM system that is likely to be installed by the major powers would be able to cope with any likely level of clandestine missiles and would thus permit the striking forces of both sides to be reduced to zero. In the process of reducing the striking force and increasing the ABM system there is a cross-over point, and Wiesner says rather optimistically that it is 'a point of considerable worry, but one which I believe can be overcome'. In view of the sorry state of disarmament prospects as a result of over-conservative judgments of military nuclear needs, it would seem extremely difficult indeed to induce nations willingly to approach the point where the relative capabilities of defence and offence are quite unclear.

The region-by-region disarmament technique is quite as useful for the solution of this problem as it is for the problem of disarmament without ABMs. The older proposals suggested dividing each country into perhaps six or more regions of roughly equal nuclear-military importance, but let us here consider the less cautious possibility of dividing each country into only two parts, the dividing lines to be chosen by the countries themselves with a view to equipartition of elements of a deterrent system but with not too unequal division of population. Let the first half to be disarmed be chosen, for example by the 'other side', and sealed off and disarmed and inspected. The host country could choose whether to leave ABM systems in operation and permit them to be inspected, or to permit them to be destroyed without detailed inspection to avoid compromising intimate design features. The uninspected half of each country would in the meantime carry the burden of deterrence.

In the inspected half of each country it would now be possible to set up the basis for a low-level deterrent, known and inspected but invulnerable to attack by the opposing low-level deterrent. Invulnerability could be arranged by one of the schemes discussed in the past, such as retaining more silos than ICBMs with occasional inspection of groups of silos in such a way that inspectors would have an accurate count of missile but not know at any one time which silos are 'loaded'. If the installed ABM systems are judged to be good enough to bring effectiveness of the low-level deterrent into question, the ABM in the inspected halves would have to be weakened to assure there being vulnerable targets sufficient for deterrence.

This done, the disarmament of the other half can proceed with greater confidence. Doing it boldly at one stroke might give rise to doubts about the possibility that one side would hold back just long enough to have a temporary superiority in ICBMs enabling it to overwhelm the relatively small low-level deterrent of the other side. One answer to such doubts is provided by an appropriate ABM system. The short-range component of an ABM system can provide a 'point defence' of the low-level deterrent ICBMs without the opponent populated targets, and a good point defence would greatly increase the number of attack missiles required for a successful first strike. Another answer is to disarm the armed half of each country by stages, dividing it in half and again in half in the process, until the last section to be disarmed is not very powerful relative to the low-level deterrent. A combination of the two answers might be the most convincing answer.

With so many years having passed with so little progress towards disarmament – nay, with so much progress away from disarmament – it seems too much to hope for the realization of such thoughts. Perhaps we should be more modest and promote an agreement between the major powers to carry out only the first step of this plan – the disarmament and inspection of one half of each country, leaving agreement on any further steps to the future. This could be regarded not only as a step hopefully to be followed by others, but it would in itself contribute to the stability of deterrence if ABMs should some day become more effective than they now appear to be. If one country should ever become so overconfident in its ABM system as to believe itself almost immune from attack, it might become too bold in its foreign policy. In this sense, ABMs may be destabilizing. Having one half of each country relatively naked would avoid this danger. Another advantage would be that this half-way disarmament plan would be an impressive act of restraint tending to persuade hesitant non-nuclear-weapon nations to accept the Non-proliferation Treaty.

In these discouraging times, disarmament seems to be a vain hope. It is right that we should discuss paths toward disarmament, that we should hope for disarmament and that we should work for disarmament. But it would be a mistake to believe that success is just around the corner. In finding in this discussion that an ABM system is not only compatible with disarmament but may even be a useful tool on the way towards disarmament, we should not conclude that this constitutes an argument for deploying ABMs. On the contrary, ABM deployment is an open encouragement of counter-measures that add to the upward spiral of a dangerous arms race the momentum of which is the principal obstacle to disarmament. We should do all in out power to try to stop ABM deployment at an early stage. But if it proceeds in spite of us, it is useful to be aware that it may be a technical aid to disarmament steps if the political will ever develops.

While we can thus find a disarmament silver lining to the ABM dark cloud, the same is not true of that other new development, multiple-warhead missiles or MIRVs as they are called. These are destabilizing to deterrence in the sense that, with the future development of sufficient accuracy, one missile can in principle destroy several opposing missiles in a first strike. This development would make more difficult the planning of a transition to disarmament and the prospect of it is one more evidence of the need to achieve at least arms limitation promptly. In view of the likelihood that the development of greater accuracy will involve tests with very long and observably multiple trajectories, a suitable subject for discussion would be whether or not it is already too late to promote a MIRV test ban.

### *Reference*

1 J. B. Wiesner, *Bulletin of the Atomic Scientists*, Jan. 1968.

## SECTION B: *Political Implications*

H. AFHELDT; C. F. BARNABY;
F. CALOGERO; J. DELBRÜCK;
J. PRAWITZ

# 1: Implications of Superpower Deployment of Anti-Ballistic Missile Systems for Third Countries, particularly for those in Europe

*Introduction*

The decision by the superpowers to deploy ABM systems has major implications for the future of International Society. Although the initial decisions for limited ABM deployment have already been taken, other important decisions must soon follow. It is the purpose of this paper to discuss certain considerations which may be relevant to these future decisions, focusing attention particularly on the secondary effect of ABM deployment. It is realized that decisions determining the extent of the ABM systems deployed by the superpowers will be mainly made on the basis of their considerations relating to their reciprocal military postures and to the internal political pressures arising out of their reciprocal military postures. It is suggested, however, that a serious effort should be made to direct the attention of decision-makers and of public opinion towards appreciating the secondary consequences throughout the world that such a decision would have, so that they can also be taken into account in weighing the pros and cons of restricting deployment.

There are many uncertainties associated with the accurate evaluation of ABM systems. International politics are, however, affected more by perceptions than facts and it is important, therefore, to assess how the peoples of the third countries are likely to perceive the deployment of ABM systems by the superpowers.

*Political Implications*

There are many implications of ABM deployment but we will only deal with some of these, namely: the destabilizing effects; the consequences for existing security alliances; the effects on arms control measures and disarmament; and the special situation created for neutral countries. The fact that the issue of the reinvigoration of the arms race is not mentioned here does not mean that its importance is underestimated; in fact, it will provide the background for most of the following considerations. It is, of course, clear that all the complex arms race implications of ABM deployment should receive careful study and analysis.

*Destabilizing Effects*

(a) *General war.* Present day superpower stability rests upon a second strike assured destruction. This stability would not be greatly disturbed by relatively large changes in offensive capabilities or, in other words, by changes in numbers, design and types of offensive missiles. It is of fundamental importance that both superpowers know that they both have, and can both maintain, the capability to strike back (assured destruction). This state of affairs is often described as a balance of terror and is, in fact, a relatively stable balance of terror. Under this balance and the necessary coexistence it entailed, there was a growing *détente*. This *détente* allowed some progress to be made in arms control and opened a real prospect of progress towards disarmament while, at the same time, maintaining stability based on deterrence. This appeared to be the only road to complete disarmament.

The present officially announced level of ABM deployment certainly will not seriously impair the present military balance.*

* The present American deployment has been described as an anti-Chinese system (McNamara, 1967) but the fact that Sprint missiles are being installed to protect missile-launching sites is hard to reconcile with this claim. China is at present a relatively small nuclear power and is likely to remain so for some time. In these circumstances it could, at most, attack only a small fraction of the more than 1000 existing American missiles. Such an attack would clearly not significantly reduce the ensuing American retaliatory blow. Therefore, the limited deterrent of a small nuclear power must, if it is to be at all credible, be entirely aimed at cities rather than missile installations. Consequently the U.S. deployment is relevant to the military balance between the superpowers.

However, the practicability of ABM deployment opens up the prospect that one or other of the superpowers might foresee a chance of attaining a dominant position in the long run. The political implication of this is to change the situation from coexistence as an obligatory state to a situation where a decision between coexistence and a competition for world dominance will again become necessary. This cannot be discussed here in greater detail although clearly this is a fundamental issue of ABM deployment.

In addition to the change in the political environment, there are other major dangers inherent in the development of ABM systems. For example, it may eventually appear to one side that it would be advantageous for it to make a first strike.

(b) *Superpower involvement in local conflicts.* The development of a really effective ABM system would make each superpower relatively invulnerable so that its involvement, even at the nuclear level, in local conflicts would be less risky and therefore more probable. Even though such a development is at present unfeasible the mere indication that the superpowers are striving towards such a posture could raise a new sense of insecurity in the other countries; the superpowers are protected by such a system and have the above-mentioned opportunity to intervene while the other powers are denied this protection, and their fear of a military intervention by the superpowers is enhanced. This sense of insecurity will inevitably result in increased armaments of all kinds in most states.

Even the ABM system actually deployed or planned might modify relative strategic postures sufficiently to affect the methods by which superpowers handle crises in the direction of less caution.

(c) *Consequnces to neutral countries.* The deployment of ABM systems directly affects neutral countries. These countries have to consider their position in a future war involving both intercontinental missiles and anti-ballistic missiles. Other countries remaining outside such a war are also considerably affected.

Firstly, the introduction of steerable missiles, multiple

individually targetable re-entry vehicles, zigzagging missiles, more powerful penetration aids, large arsenals of offensive missiles, etc., will increase the probability that some missiles might get out of control and land nuclear warheads in a haphazard manner, possibly on neutral soil.

Secondly, the light flash from nuclear explosions in space may cause casualties in neutral countries, either because of retinal burns in the eyes of persons looking towards the sky or by temporary flash blindness. This danger may seem small compared to the effects of general war on the countries directly involved but countries actually facing this potential threat may be induced to take counter-measures.

Some of these countries may develop electronic devices to counter the threat to them and this raises several questions. For example, there is the question of whether such measures comply with the status of neutrality or non-alignment. There is also the question of how far in space the territorial atmosphere extends.

The countries concerned must decide whether or not to install electronic techniques designed to steer away, or render inert, missiles passing over their territory. If they do so, the probability that the deflected or inert missiles would hit some other country is increased. These issues clearly involve not only questions of the status of neutral countries in international law but also serious political problems for many third countries.

### *Effect on Existing Security Alliances*

Although the aim here is to concentrate on specific European alliances many of the following conclusions are also relevant to all existing security alliances.

There are many factors which will determine the future of European alliances, of which ABM deployment is one. It is, however, an important factor which has tended to be seriously underestimated in European debates on the future of the security alliances. As was indicated above it has an important determining influence on the stability of the strategic level between the superpowers and, since European security today depends to a large extent upon such stability, the European situation is clearly affected by any substantial change in this level

At first sight the argument that the defence of the superpowers by ABMs makes their commitment to European allies more credible appears convincing. However, the reliance on stability based on the inability of a potential attacking power to calculate the victim's response (stability by uncertainty) had, for the European states, one advantage: this uncertainty gave them the assurance that the interest of the superpowers was also their own interest; the superpowers and the Europeans were equally anxious to prevent a conflict in Europe because the possibility of escalation was perceived to be high. Until now such an escalation was equally dangerous for both the superpowers and the European powers.

An even more important consequence of ABM deployment arises from a psychological and political factor: namely, the feeling (justified or nor) that the superpowers are protected by the defence shield provided by their ABM systems whilst Europe remains undefended. This feeling is an emotional one and largely independent of the actual effectiveness of ABM systems. Possible results of this range from fuelling nationalistic feelings of frustration, insecurity and resentment against the superpowers to a constructive tendency causing non-nuclear European states to move closer together, thereby removing Europe from direct superpower confrontation. Although this latter tendency is also reinforced by other objectives forces tending to produce European collaboration it is not possible to predict which of these tendencies will prevail. It is also, however, not impossible that some governments or peoples may argue that, for the sake of keeping the alliances intact and for their own defence, ABM systems should be extended to cover Europe. Not only is this probably technically unfeasible but it would also deepen the rift between East and West, thereby re-enforcing the confrontation at all levels.

Similar considerations apply to arguments that Europeans should develop their own ABM system independently of the superpowers. Since even the ABM systems of the superpowers, in the present state of ABM development, will not result in a significantly improved security for them, any European system has even less chance of being effective. However, as happened within the superpowers, this is not a sufficient reason to prevent political pressure for it from arising in Europe, particularly in view of the deployment by the superpowers.

*The Non-proliferation Treaty and the Test Ban*

The decisions already taken on ABM deployment have significantly influenced, in a negative way, the current negotiations for a non-proliferation treaty. This influence arises in three main ways.

Firstly, there is the general reaction in the non-nuclear countries to the unfortunate example set by the superpowers. When faced with the choice between restraint in weapons development and deployment on the one hand and the pressures of the military and unenlightened public opinion for such development and deployment, on the other hand, the superpowers have consistently given the impression of yielding to the latter. Many third countries feel that restraint produces a *détente* between the superpowers, an easing of international tensions and progress in arms control, and consequently some non-nuclear countries resent this aspect of superpower behaviour – especially when the resulting weapons deployment makes little sense on strategic grounds. The hesitation to sign a non-proliferation treaty is a good example of the way in which the resentment is demonstrated.

Secondly, non-nuclear countries have argued strongly in favour of a 'balanced' treaty which would oblige the nuclear powers to undertake positive measures. Although the partial test has been appreciated by the non-nuclear powers the latter feel that it is insufficient on its own and that the new measures should go beyond this; a comprehensive test ban agreement, a cut-off in the production of fissionable material for weapon production, and a freeze in the deployment of nuclear delivery vehicles are measures that have been suggested. While the nuclear powers understand this view they claim that more time is necessary to negotiate the measures suggested; they also argue that the present absence of agreement on these measures should not delay a non-proliferation treaty and that, on the contrary, a non-proliferation agreement now would facilitate ater agreement on more comprehensive measures. It is the credibility of the latter statement that is seriously weakened by a procurement of ABM systems since such procurement is related (or is perceived to be related) to a need for weapon testing. This will, in turn, increase the already serious hesitation from some countries, in particular a few important near-nuclear countries, to sign a non-proliferation treaty. However, this unfortunate aspect of ABM deployment could be counteracted by convincing measures taken by the nuclear powers. An example of such a measure would be a credible assurance, e.g. a freeze in nuclear delivery vehicles, that a more extensive ABM deployment will not ensue. Although the above arguments carry great weight in some sections of opinion in some non-nuclear countries they should not be pressed to a point which would make an early solution of the non-proliferation problem impossible.

A third consequence of ABM deployment on the prospects for a non-proliferation treaty is related to the possibility that a growing political demand may develop for European ABM systems if the superpowers decide to extend their systems. Because of the short warning times involved, a European system would require a revision of the well-established compromise on nuclear decision-making within alliances. Therefore, pressure to keep open an option for European ABM systems will push the non-proliferation issue back to the point it was at three years ago. Even an extensive debate on the issue of ABM systems for Europe will probably delay the negotiations for a non-proliferation treaty.

It has been argued, however, that a widened gap between the superpowers and prospective new nuclear powers is an obvious hindrance to proliferation because a relatively small offensive force will no longer deter an enemy equipped with an ABM system. This clearly does not apply to many of the non-nuclear countries whose potential enemies are either non-nuclear or potential small nuclear powers. At first sight, it might appear that the argument is valid for those non-nuclear countries interested in obtaining a minimum deterrence against the superpowers. However, if the respective nuclear power is a neighbour or a near-neighbour of the superpower the argument is not necessarily valid. In such cases, the power concerned would probably arm itself with short and medium range missiles rather than intercontinental ballistic missiles. An ABM system deployed against intercontinental missiles is of little value against low-flying short and medium range missiles* and does not, therefore, significantly alter the credibility of the deterrent of a small nuclear power armed with such weapons. The strategic deterrent of a small nuclear power against a superpower is based upon the assumption that the superpower would not be willing to accept the loss of even a few of its cities if its only gain is the destroyed territory of the small power. Therefore, it can be argued that the small power has some deterrence if it can produce some damage to the superpower; the superpower, even defended by an ABM system could not exclude all damage. The deterrence of a superpower against another superpower is based on the different doctrine of 'assured destruction' in which each superpower perceives that the other will accept heavy losses for the sake of 'winning' a war. Furthermore, the tactical use of nuclear weapons in defence of territory against an attack by a superpower is not affected by an ABM system in the homeland of the superpower. But it should be emphasized that it can also be argued that these remarks rest on rather doubtful premises. In fact, the strategic arguments which have been given to justify the need to go nuclear are not very convincing; they justify the suspicion that they are concocted *a posteriori* to provide some intellectual respectability for a decision which is largely based on political and emotional motivations. This also applies to the ABM deployment undertaken by the superpowers; the most important question is the political and emotional impact of this deployment rather than its actual strategic relevance. The perception that the advent of ABMs has 'raised the entrance fee into the nuclear club' might have some anti-proliferatory value; this effect may, however, be com-

* Of course, if a superpower were really interested in neutralizing the potential threat of a minor nuclear power close to its border, it might modify the design of its ABM system so as to increase its effectiveness against medium range missiles.

pletely reversed for the reasons stated above (the 'bad example' set by the superpowers, increased resentment and frustration, and decrease of the *détente*), and, in any case, it is of limited importance.

So far as the test ban is concerned, it is briefly noted that the procurement of ABM systems is likely to further postpone the conclusion of a comprehensive test ban foreseen in the preamble of the Moscow Treaty. In fact, ABM development has significantly increased the need for underground testing;* there is also strong pressure for testing in space.

*Conclusions*

Until recently there were some indications that man was beginning to cope with the conditions brought about in international society by the technological revolution. It appeared that, at last, some progress was being made towards arms control. Governments seemed increasingly unwilling to pursue their goals by the use of military power.† There were some signs that even the most powerful states were becoming willing to modify their goals to accommodate peaceful change in international relations.

All states have urgent and serious internal problems and many, including the superpowers, are under strong pressure from their populations to concentrate their energies and resources upon them. It looked as though the superpowers genuinely desired a period of *détente* and nuclear stalemate in which to attempt to solve these problems. Should the superpowers re-engage in an armaments contest the fragile progress that man has made will be quickly reversed.

It is for this reason that future decisions on ABM deployment are so important. Man has to be convinced that he must control the rapid advances in technology instead of accepting the

* A few underground tests have accidentally released radioactivity which has spread outside the testing country in direct contradiction to the Moscow Treaty.

† Military involvements by the superpowers should have demonstrated to the superpower concerned that this type of involvement no longer achieves the desired objectives.

dictatorship of technology. ABM deployment is a glaring example of policies shaped by weapons technology. Technology makes these weapons available; because they are available political pressures build up within states for their acquisition; political leaders succumb to this pressure and invent policies to justify the acquisition of the weapons. Once the weapons are acquired strong pressures develop for their large scale deployment even in the absence of sensible strategic arguments. Judging by past experience, therefore, the present thin ABM systems will almost certainly be thickened in future. The consequences of this have been described above; the nuclear stability of the superpowers will be eroded, the arms race will get out of control and the effects discussed in this paper will follow. The results of this situation would be very serious for all members of international society. In our view the extension of the ABM systems will only be avoided if the leaders of the superpowers intervene.

It follows that the leaders of states, especially of the superpowers, should exercise their leadership in the direction of arms control and peaceful adjustment to change in international society and should educate their citizens in an attempt to break this dangerous spiral. Unless these leaders show self-discipline in providing their states with the military power that technology makes available to them the mood of frustration that will inevitably follow might well be irreversible. International tension will increase, possibly to breaking point.

Part of the role of European states is to exert influence on the leaders of their respective blocs – the way they should use this influence is clear.

B. T. FELD

# 2: Implications for Other Nations – particularly in Europe – of Superpower Anti-Ballistic Missile Deployment

So much has already been written on the consequences of the introduction of ABMs into the nuclear deterrence formula – both with regard to its destabilizing effects on the nuclear arms race and to the (in my opinion, highly hypothetical) possibilities that ABMs might be used as a device to provide some stability at a relatively lower and mutally acceptable level of offensive missile deployment – that it would be presumptuous to try to add to this discussion. Instead the remarks which follow are concerned with the implications of recent developments on the possible actions of nations other than the nuclear superpowers, and with possible circumstances under which concerned nations might influence the decisions of the superpowers on future ABM deployment.

The following considerations appear to be relevant to such a discussion of the implications of ABMs.

1. Any ABM deployed is likely to be appreciably less effective than claimed by its proponents.

First, the necessary time-lag between the decision to deploy and the deployment of a working system adds to the intrinsic advantages of the offence for the development of countermeasures and the accumulation of sufficient numbers of offensive missiles to saturate the defence. Second, the Test Ban Treaty and the Non-proliferation Treaty both render it extremely difficult, if not impossible, to test adequately the extremely complex ABM systems under consideration.

Furthermore, the obvious advantages to the superpowers of the continuation in effect of these treaties, especially the latter, render it extremely unlikely that they will be violated or denounced for the marginal advantage of testing ABM components. This conclusion is strengthened by the tendency of the military in both superpowers, in their anxiety to sell the systems, generally to over-estimate by a large factor their effectiveness. On the other hand, the reaction of either of the superpowers, to ABM deployment by the other, is likely to be based on the most optimistic military estimates of their effectiveness. It is this factor which generates the 'mad momentum' of the nuclear arms race, and renders so unconvincing the arguments of the 'limited' ABM proponents for its potentially stabilizing aspects.

2. The recent decisions, both in the Soviet Union and in the United States, to embark on limited ABM deployment, are based much more on political and psychological than on convincing technical arguments.

As a corollary, such decisions can be reversed, on political grounds, without seriously affecting nuclear deterrence and its stability, provided that the reversals are not too long delayed. In this regard, the reaction of the ABM proponents in the United States Senate to the recent Soviet announcement of willingness to engage in discussions on limitation of both defensive and offensive nuclear delivery systems – that this was an effect aimed at by the decision of the United States for a limited ABM deployment – provides both a hope, that mutual agreement on limitations is still possible, and a warning that failure to arrive at such agreements will only intensify the pressures, at least in the United States for a massive expansion of the currently limited ABM deployment.

3. ABMs would be no more prohibitively expensive for other countries than embarking on the achievement of a significant nuclear delivery capability.

If the United States can afford to spend about $30 billion a year on the Vietnam war, a number of other technologically developed countries can certainly afford the few billions required to provide themselves with the psychological lift of a limited nuclear capability, either offensive or defensive. The coming into being of the Non-proliferation Treaty applies a brake on such possibilities, but this brake is at best temporary, since the treaty is unlikely to remain in effect if the nuclear arms race between the superpowers is not brought under control. The argument, that ABMs require an effective civil defence programme, especially if it were attempted by the densely populated European nations, is only serious in so far as ABMs are technically and not politically motivated, which would not apply to a European ABM deployment. Since civil defence is a joke, the joke can be equally effective with a token effort, costing relatively little, as with a serious effort, impossibly expensive if at all feasible. It is probably the case that for an area as densely populated as Europe, in which all the nations (in the East as well as in the West) are so close to the bases of their potential 'enemies', ABMs make even less sense than for the superpowers – which is to say that it makes no sense at all.

4. ABMs are but one aspect of the destabilizing effect on the continuing arms race of new technological developments, which include MIRVs, FOBS, BWs (biological weapons) and CWs (chemical weapons).

Some of these, even if apparently more remote, pose greater dangers in their destabilizing effects. In particular, the development and deployment of MIRVs by the superpowers presents an imminent danger that one or the other may become convinced that it has achieved the possibility of a successful pre-emptive first strike; such a conviction (even if illusory, as it probably would be) could be more upsetting to the stability of international relations than ABMs. However, as far as the smaller nations are concerned, the development of biological and chemical weapons, and their dissemination, poses a far greater danger to the stability of an already shaky international order (cf. the Middle East).

5. For some time to come, China will not pose any serious military threat to the rest of the world.

If this assertion – based on the continuing preoccupation of the Chinese leadership with internal problems, as well as on the realism which this leadership has previously demonstrated with respect to external military adventures – is correct, there is appreciable time available for the rest of the world to work out a system of arms control into which China can eventually be incorporated. In any event, the problem of China and arms control need not be faced in our immediate considerations.

How does all this affect the decisions which European nations, and other near-nuclear countries, must make in response to ABM deployment by the superpowers? Probably the first, and most important, conclusion is that the most sensible decision on their part would be not to respond at all. There is no immutable law of international relations which requires foolishness to beget foolishness – although history provides too many examples of such behaviour. However, in this case, self-interest is in strong opposition to historical precedent; the European nations have many more constructive needs for their resources than to become involved in the fools' game of the nuclear arms race.

But, unfortunately, true self-interest has seldom been the prevailing motivation for national action and reaction. With the growing prosperity and economic and political independence of many European nations, there will be a tendency to emulate the French decision to construct an independent 'force de frappe' (or even more) which, technically unproductive as it has been, has nevertheless provided France with certain psychological benefits.

However, the present situation does confer on the non-nuclear nations a possibility, as well as an opportunity, to play a much more positive role. For if the nuclear arms race is to be reversed – and this would be to the demonstrable advantage of all countries – it will only occur if there are unrelenting pressures on the nuclear powers, exerted by the nations which have it within their possibility to emulate their irrational behaviour, to take positive steps to curb their competition in nuclear arms accumulation. And such pressure can only be effectively exerted if the non-nuclear nations continue to refrain from entering into the competition for accumulation of weapons of mass destruction. Enlightened self-interest would dictate such a course, even if narrow nationalist tendencies oppose it. But the adoption of a course of rational action requires that rational arguments shall be heard, understood and accepted. Therein lies the challenge now confronting the scientific community.

K. MORTENSEN

# 3: Some Observations on the Political Implications of Anti-Ballistic Missile Systems

A question which inevitably will arise in connection with an evaluation of certain political implications of deployment of ABM systems is this: Do these weapons have a stabilizing or destabilizing effect on the international situation? Here, already, lies the first difficulty, which illustrates how intricate the ABM problem is. The two powers which have begun, or have resolved to begin, ABM deployment will presumably judge this question from different angles, mainly because of their different strategic doctrines and their different demographic and industrial concentrations. It cannot be ruled out, however, that the powers – notwithstanding their dissimilar structural patterns and fundamental views – may agree on the expediency of reconsidering their own goals and arguments as well as those of the other party. This seems to be borne out by the most recent developments as they have manifested themselves in President Johnson's statement of 1 July 1968, and in the Soviet Government's memorandum of the same date. The possibility that world opinion concerning the influence of ABM systems on the international situation, has contributed to these developments should not, however, be ignored.

The limited deployment which has been initiated and/or decided upon in the Soviet Union and the United States can hardly, in itself, be a destabilizing factor. Indeed, some might even think that it will have a limited stabilizing effect. If started, a deployment may, however, make the opponent feel uncertain, particularly about its scope and effectiveness, and it may engender fears that the evolvement of these types of weapon may result in a technological breakthrough that may change the strategic picture, at least as seen through the eyes of the opposite party. It is, today, fully justifiable to ask whether the build-up of the existing over-dimensioned arsenals of ICBMs was not, in fact, based on miscalculations of the other party's capacity. This recognition was manifest in the speech by the American Secretary of Defense, Mr McNamara, at San Francisco on 18 September 1967: 'The blunt fact remains that if we had had more accurate information about planned Soviet strategic forces, we simply would not have needed to build as large a nuclear arsenal as we have today.' This miscalculation may be repeated in the ABM case.

It would seem appropriate in this context to recall Professor Niels Bohr's Open Letter of 9 June 1950 to the United Nations about an open world. The letter quotes a memorandum of 17 May 1948, which Niels Bohr submitted to the Foreign Secretary of the United States in June of the same year. It says, in part:

> In the years which have passed since the war, the divergencies in outlook have manifested themselves ever more clearly and a most desperate feature of the present situation is the extent to which the barring of intercourse has led to distortion of facts and motives, resulting in increasing distrust and suspicion between nations and even between groups within many nations. Under these circumstances the hopes embodied in the establishment of the United Nations Organization have met with repeated great disappointments and, in particular, it has not been possible to obtain consent as regards control of atomic energy armaments. In this situation with deepening cleavage between nations and with spreading anxiety for the future, it would seem that the turning of the trend of events requires that a great issue be raised, suited to invoke the highest aspirations of mankind. Here it appears that the stand for an open world, with unhampered opportunities for common enlightenment and mutual understanding, must form the background for such an issue. Surely, respect and goodwill between nations cannot endure without free access to information about all aspects of life in every country.

These thoughts, presented twenty years ago, are equally

valid today. But, regrettably, it seems just as difficult now as then to have them generally accepted.

However, the demand for openness will remain on the agenda. The negotiations which are now expected to be opened between the Soviet Union and the United States for cut-backs in stockpiles of not only ABMs but also offensive missiles, will inevitably bring up the question of verification. How could a reduction of nuclear stockpiles be considered unless we know the initial positions, and how would it ever be possible to verify observance of a treaty on such reductions without some degree of recognition of the requirement for openness? If a treaty is to be based upon the principle of unilateral verification, suspicion will be allowed to live on, to the detriment of the disarmament cause. In any event it would be wise not to pin too high hopes on the coming negotiations; they are bound to be protracted and to become extremely difficult. It is gratifying that these negotiations will be started, but as long as they are under way the ABM problem will remain acute. A reciprocal moratorium, while negotiations are taking place, is not very likely, seeing that it will hardly be possible to consider ABM systems in isolation from offensive missiles.

But even when leaving out of account any major miscalculation on either side with regard to the defence machinery of the other, it seems more than questionable, judging by present knowledge, whether ABM systems can have a stabilizing effect. If ABM systems were used to defend launching sites for offensive missiles they might, theoretically, enhance the credibility of the guarantees undertaken by the country concerned, but the interest in the use of these defensive weapons seems so far to be focused on protection of population centres and industrial centres. In addition, it will hardly be feasible at present to mount a 100 per cent safe defence against missiles, and penetration of an ABM system could thus be achieved through a major offensive effort. The only acceptable justification for deployment would be that an ABM system could offset an inferiority in regard to offensive weapons, but accessible data make it abundantly clear that from a cost/effectiveness point of view, greater deterrence will be attained by concentrating on the evolvement of offensive weapons. On the other hand it would undoubtedly be wishful thinking to imagine that questions of importance to national security are settled by financial considerations. However, it seems to be a realistic judgment, although couched in rather general words, that recourse on both sides to deployment of major ABM systems predictably would end at a stalemate on a higher, costlier, and perhaps more dangerous level of nuclear stand-off.

However, this evaluation will probably have to be modified if we turn from the relations between the nuclear super powers to the relations between the big and small nuclear powers. In the latter relations, ABM systems may be more effective. The official American motivation for deployment of ABM systems is, in fact, to set up a defence against China. The decision of the United States has undoubtedly also been strongly influenced by psychological factors. Greater defence capabilities *vis-à-vis* smaller nuclear powers will strengthen the hegemony of the superpowers and may make the lesser powers feel even more helpless towards and distant from the superpowers. The recent discussions of the Non-proliferation Treaty at the resumed 22nd General Assembly of the United Nations reflected growing misgivings about the idea of added discrimination in the field of national security. Whether these reactions were justified or not is beside the point in this connection. The fact is that they were there.

In relation to China the deployment of ABM systems may accelerate a protection which may be considered acceptable. The superpowers will thus, for a time, be able to keep a Chinese nuclear threat in check. But this deployment is not likely to be made for the sake of China alone. It is, admittedly, generally understood that China's development of nuclear weapons has made faster progress than envisaged; but it should not, at a time when Chinese possession of effective ICBMs is still several years off, be necessary to embark upon a Chinese-oriented ABM system. Conversely, China is likely, in the longer run, to develop sophisticated nuclear weapons against which a thinly deployed ABM system would still not be sufficient. In relation to China, therefore, the deployment of ABMs must be regarded as an attempt to gain time. But this will not solve the longer-term problem of drawing China into the international debate. It is more likely to help maintain China's isolation even if that is partly of China's own choice. This policy, if continued, will lead to a further escalation of the arms race.

In theory, one positive element of an ABM deployment could be that it would induce prospective nuclear powers to refrain from acquiring nuclear weapons and thus serve as a guarantee against proliferation of such weapons. The value of this argument is, however, very doubtful because prospective nuclear powers would not take their decisions about nuclear weapons in the light of their relations with the nuclear superpowers; their decisions would be based on their relations with neighbouring states which are not nuclear powers or which have no ABM systems.

In the field of disarmament, a deployment of ABMs would, incidentally, detract from the possibilities of achieving arms control:

> It would lead to a growing interest in and probably also a persistent need for continued nuclear tests. The prospects of reaching agreement on a complete nuclear test ban would probably be decisively weakened.
>
> In a situation where the United States and the Soviet Union feel compelled to augment and improve their missiles capacity, the chance of realizing the proposals for discontinuation of the production of fissionable materials is probably nil.

If ABM systems are further developed in the United States and the Soviet Union and if the superpowers continue along the road they have taken, there is a risk that a clash between great powers would, at least in the initial stage, result in nuclear attacks against those allies of the great powers which had no ABMs. If this hypothesis is correct, such allies are likely to claim the same protection as the protecting state. Such claims may be raised especially in Central Europe where there are heavy concentrations of people and industry. But it is most unlikely that any reasonable protection can be obtained against nuclear attacks in Central Europe. The range of available delivery vehicles would be much greater in that area than it would be for an exchange of nuclear missiles between the superpowers, especially in the forms of low-flying aircraft, medium-range missiles, nuclear artillery, etc. It should also be borne in mind that there are so many centres in Central Europe that protection would, at best, have to be selective, also for economic reasons, but any areas selected for protection would no doubt attract the enemy's special attention – quite apart from the great political difficulties that would arise in any such selection of centres to be protected. So, instead of affording protection, a deployment of ABM systems would be more likely to attract enemy attacks. Finally, the adoption of a non-proliferation treaty would prevent deployment of ABM systems because the operation of such systems would probably depend on their being armed with nuclear warheads. Joint European solutions would also be precluded by a non-proliferation treaty; moreover, there is nothing in the present situation in Europe that would seem to motivate such a solution. What is left, then, appears to be a growing feeling of Europe being the hostage of the superpowers – a role that some countries may find more acceptable than others, but most of them will find that it bears no reasonable relation to their achievements in non-military fields. It is certainly very difficult to say what the reactions are going to be, but it is by no means unthinkable that ABM build-ups in the United States and the Soviet Union may generate psychological setbacks in relation to the superpowers and thus have adverse effects on *détente* in Europe.

Humanity has long been living in fear of nuclear weapons. It may, therefore, seem paradoxical that the development of weapons that can afford some protection against nuclear attack may cause additional fears. It should be borne in mind, however, that these defensive weapons are likely to be available only to the highly developed nuclear powers. So far, humanity has had to share the nuclear horror, but the reaction stems from the fact that the very powers which hold these all-destroying weapons are also – and exclusively – the powers that hold defensive weapons against them. In summary, the destabilizing effect of ABM systems is therefore likely to be greater in the field of psychology than it is in military strategy. So, if we realize the difficulty of controlling psychological reactions we shall get closer to the core of ABM problems.

# SECTION C: *Strategic Implications*

F. A. LONG

# 1: The Impact of Anti-Ballistic Missile Deployment on the Uncertainties of Strategic Balance

### *Introduction*

Uncertainty is a characteristic fact of life for all aspects of strategic balance among nations but it is particularly characteristic of the weapons systems that are incorporated into modern military forces. There are several reasons for this. A prime cause of uncertainty is the rapid rate of technological change. This makes it hard to be sure that a given military posture is currently optimal. It even makes it difficult to ascertain when a weapon (or a strategy for its use) has become obsolete. Equally, the rapid rate of technological change makes it difficult to anticipate the trends on weapons or their effectiveness.

A second cause of uncertainty, which also stems from the growth of technology is the wide spectrum of system choices and weapons designs which are available to the planner. One can choose to emphasize artillery, or unguided rockets or aerial bombardment; one can give strategic nuclear responsibility to submarine-launched missile forces, to silo-launched missiles; one can even consider assigning strategic deterrence to biological weapons systems.

A third reason for uncertainty is the enormous complexity of modern weapons systems, combined with the frequent requirement for very rapid response. Modern aircraft are themselves very complex and require elaborate maintenance systems to keep them in a state of readiness; response time for their use is, however, usually a matter of hours. Rocket-borne nuclear weapons systems are equally complex and yet, at least for land-based missile forces, their command and control must permit response times within tens of minutes. It is virtually unavoidable that very real uncertainty will exist as to whether these complex and infrequently used systems will in fact respond rapidly, and will perform reliably in the very small reaction times that can characterize possible emergencies. This problem is exacerbated by the fact that realistic proof tests of these weapons are exceedingly difficult to accomplish, it being especially difficult to analyse the spectrum of possible emergencies and then simulate them appropriately. For ballistic missile defence systems, proof testing is further inhibited by the provision of the Moscow Treaty which bans nuclear weapons testing in the atmosphere. Considerable uncertainty will remain as an almost inescapable characteristic of these systems.

The question which this chapter addresses is, in view of these types of uncertainty, what will be the probable impact of ABM deployment on the strategic balance which now exists in the world? Since the emphasis will be on the nuclear balance some preliminary comments on the current situation appear in order.

For some years the principal component on the world-wide strategic balance has been the mutual nuclear deterrence which exists between the United States and the Soviet Union. This deterrence principally involves large forces of rocket-borne nuclear armed intercontinental missiles which are either housed in underground missile 'silos' or are mounted on sea-going vessels. Target accuracy is probably in the order of one mile at the extreme range of from 5000 to 10000 miles. Equipped with, for example, one megaton warheads this accuracy is sufficient to give a high probability that a targeted city will be hit and suffer very great damage. An important characteristic of the nuclear deterrent systems is that on each side the combinations of missile numbers and launch site vulnerability is such that enough missiles will survive a first strike to ensure vast destruction when they are used in a retaliatory second strike mode.

The existence of a continuing mutual deterrence depends partly on the fact that nuclear weapons, with their awesome capacity for death and destruction, appear to have shown themselves to be superior weapons for retaliatory use but inferior ones for political and diplomatic use. This low effectiveness of nuclear weapons in political situations is particularly evident when two opposing nations have them; they are also of comparatively low political utility when a large nuclear-armed nation opposes a small non-nuclear nation since the moral and psychological inhibitions against use of nuclear weapons remains high. Hence the political impact of a threat to use these weapons is not intolerably great and the pressures towards obtaining a 'first strike' capability can be kept in bounds. At the same time the destructive capability, in a retaliatory strike, of a reasonably non-vulnerable nuclear weapons force is so tremendous that the deterrence value of the force can be high. Furthermore, subtle characteristics of the nuclear deterrence force are not important in maintaining its utility as a deterrent and even numbers of weapons is of only moderate import if indeed only a retaliatory use is sensible. Because of these facts it has been possible for the United States and Soviet Union to speak of a 'stable nuclear deterrence', implying that the uncertainties in the mutual assessment of each other's nuclear forces was not sufficient to lead either the United States or the Soviet Union into major new military or political initiatives.

One can argue that much of this 'stability' is only psychological, i.e. not consistent with the technical facts, but even a psychologically based stability can be real and important. We must now ask whether technical or other developments seriously threaten this stability and, if so, what consequences are likely to follow.

### *Technical Uncertainties*

Since the establishment of the current position of strategic balance has principally depended on the utilization of systems which stem directly from recent technical developments in nuclear weapons, in long-range rockets and in rocket guidance, it is unreal not to expect threats to this balance to enter from further changes in technology. As a minimum one could expect bombs and rockets to become more efficient and guidance accuracy to improve. But since the efficacy of mutual deterrence has seemed rather insensitive to small changes of this character, this kind of evolution has not carried the air of a serious threat to stability. In contrast, however, the development of a truly effective defence against long-range ballistic missiles carries the potential of destroying the effectiveness of a nuclear retaliatory force and hence of seriously upsetting the strategic balance. Furthermore, it is highly doubtful whether a truly effective defence is even feasible, i.e. one runs the risk of a weakening of the strategic balance with no counterbalancing gained in individual nation security. It is in this context that the deployment of ABMs by the Soviet Union and the projected deployment by the United States must be viewed.

Technically an ABM system is highly complex. It involves nuclear-armed missiles with very short reaction times and a variety of highly sophisticated radars. The entire system is tightly linked to advanced computers which give the necessary speedy analysis and system control. Time for decision for the system is necessarily very short; only a very few minutes can elapse between initial radar detection and the launch of the defending forces if missile interception is to be successful. In view of these complexities and the tactical problems discussed below, there will necessarily be very great uncertainties in estimates of the reliability of an ABM system. The offensive forces arranged against an ABM system need to do their planning under the assumption that its reliability is high: the defenders, however, will probably make the conservative assumption that their own reliability is low and hence tend to overbuild.

There will be even greater uncertainty in assessing the effectiveness of an ABM system since this depends on the properties of both the ABM system itself and the attacking forces. The attacker can equip his forces with decoys and other penetration aids in an effort to counter ABMs. He can try to shield or otherwise modify his nuclear warheads so they can survive an ABM attack. He can modify his attack strategy in an attempt to saturate certain components of an ABM system or to induce nuclear 'blackout' of the defending radars. The

ABM designer can in turn respond to these various initiatives or to the threat of them. He may deploy more radars and ABMs; he may otherwise improve the system capabilities; he may introduce a mixed system of short-range and long-range interceptors. The almost inescapable result of these possible activities is a high degree of uncertainty in the effectiveness of both the ABM system and the attacking missile system. Once again the probable response will be a 'defence conservative' one. The ABM system group will be conservative in its assessment of the ABM effectiveness and will call for increased defence levels. Similarly the attacker will be conservative in his estimates of the effectiveness of his countermeasures and will urge more offensive missiles and more penetration aids.

What are the consequences of these technical uncertainties? The one which is implied by the last remarks is obvious enough, a strong pressure towards acceleration of the arms race. A reciprocal action-reaction behaviour is inherent in the entire arms deployment business. But uncertainties of the type involved here will greatly exaggerate the problem.

A second consequence which is discussed in the next section is changes in strategy and perhaps also in diplomacy.

A third and rather curious possible consequence of these technical uncertainties is a movement towards symmetry in military forces. If the Soviet Union deploys ABMs, there will be a tendency for the United States to do the same. If the United States develops a Polaris system the pressures on the Soviet Union to do likewise are apparently very large. At first glance this trend towards duplicating systems is a surprising behaviour since, *a priori*, one might expect that different local conditions, e.g. land power *vs.* sea power, to lead the United States and the Soviet Union in somewhat different directions. Actually the tendency towards duplication makes a certain amount of sense. Since the same science and much of the same technology is available to all, each country can assume that, given earnest effort, it can develop whatever the other can. Hence the conservative procedure is to assume that whenever another country develops a given system it does so for technologically interesting reasons and that, therefore, it makes good sense to go down the same path. It is perhaps also true that the common world-wide growth in science, and to a lesser but important extent in technology, causes opportunities for new weapons systems to become visible to the developed nations more or less simultaneously. Whatever the reasons the significant consequence is that by having systems of similar and mutually understood characteristics the uncertainty in analysing their use and especially their deterrence features is reduced to an acceptable level.

*Strategic and Political Uncertainties*

The development of ABM systems and, in response, of penetration aid systems of various sorts will suggest a variety of strategic and political responses, many of which will be a direct consequence of the technical uncertainties. A minimum response will be a modification in the targeting of the nuclear deterrence forces. Assuming that the desire is to maintain some minimum level of assured destruction it will be sensible to aim the deterrence missile forces towards less heavily defended targets. It may also appear persuasive to shift from direct attack of targets to indirect attack, e.g. by nuclear fall-out.

There will also be pressures for the procurement of new weapons systems in an attempt to maintain deterrence. FOBS may be one such response. Very low flying cruise missiles might be a second. A different and more perturbing possible consequence is that a strong ABM system may lead its possessor to contemplate a 'first strike' with his nuclear missile forces. (Against this is the argument that a conservative posture will lead a nation to derogate its own ABM capabilities and hence be very hesitant to carry our programmes which, for success, postulate fully effective ABMs.) Alternatively it has been argued that, with or without ABMs, a nation which is vulnerable to a first strike can maintain deterrence by a doomsday-like posture wherein it threatens to launch its entire missile force at the first instant of radar detection of a missile attack, i.e. before its forces can be destroyed. It remains doubtful, as between the United States and the Soviet Union, whether threat of such an overwhelming response will carry the air of credibility that a deterrence posture seems to require. On the other hand this might be a much more plausible response to a superpower from a smaller power equipped with only a minimal force of nuclear delivery systems.

A very different, essentially political, response to a large ABM deployment is to shift to some degree the focus of deterrence forces from the principal enemy to its allies. Thus in response to a large deployment of ABMs by the United States, the Soviet Union might increasingly direct its deterrence forces towards Western Europe. It is not clear whether on balance, this would be an effective strategy since the effect would be to diminish the strength of all alliances to the superpowers.

An increase in military and political uncertainty arising from extensive ABM deployment and from the probable countermeasures to it could spell the end to the growth of any significant *détente* between the United States and the Soviet Union. For a significant relaxation of tensions between two rival nations there must not only be clear perception of the mutual benefits of *détente* (and this does appear to exist between the United States and the Soviet Union) but a psychological climate which permits new foreign policy directions and new peace-oriented bilateral arrangements to arise and gain acceptance. In a word, an atmosphere of trust must be developed. But it is just this mutual trust which tends to be the first victim of the uncertainties of an offensive-defensive arms race.

This problem is particularly acute in a country which, like the United States, relies on open discussions and debate to develop support for expensive or controversial projects. The kinds of argument which will almost of necessity be used to gain support for an expensive weapons system will very often be the kinds which will raise suspicions of Soviet motives and in other ways work against the development of an atmosphere of trust and good will.

### *Implications for Arms Control and Disarmament*

The deployment of extensive ABM systems by the United States and the Soviet Union will almost surely increase the problems of obtaining significant arms control and disarmament. In a general sense this is only a restatement of the conclusions that an atmosphere of uncertainty and suspicion is an almost inevitable concomitant of an arms race and that the obtaining of all forms of agreements will therefore become more difficult. However, there are additional specific effects which will enter. In the negotiation of the proposed Non-proliferation Treaty it has steadily been pointed out that acceptance of this by the non-nuclear nations is much less likely if an increased Soviet–American arms race is underway. And even if the treaty is widely accepted, its lifetime may be short if the current nuclear powers maintain or increase their reliance on nuclear weapons. As a further specific example, a comprehensive test ban will be much harder to obtain if large commitments to ABM systems are being made. Serious reliance on ABMs carries with it strong pressures for continued underground testing of nuclear weapons to obtain new or modified weapons in response to changing requirements. Since pressures for 'realistic' atmospheric nuclear tests will undoubtedly increase as ABM systems are developed, even the Partial Test Ban Treaty may come into jeopardy.

But assuming the United States and the Soviet Union do begin serious negotiations towards a reversal of the nuclear arms build-up, the existence of extensive ABMs will be a serious complication in obtaining agreements. As one works towards a freeze and cutback in nuclear weapons systems, the evaluation of the 'relative worth' of different sorts and sizes of nuclear weapons carriers – ICBMs, MRBMs, SLBMs, aircraft – will not be easy. In fact, there is good reason to believe that only a very simple agreement will be negotiable. In light of this, the uncertainties from ABM deployment – uncertainties which relate both to the ABM systems themselves and the effectiveness of the missile forces for which they are the counter – will most surely make the development of a simple agreement more difficult.

### *Impact on European Nations*

The added uncertainties which will confront the United States and the Soviet Union as a consequence of extensive ABM deployment will have their impact on the European countries also. The principal stresses will probably enter in strategic and political spheres and on long-range military planning.

The immediate problem for the non-nuclear Western European nations will be whether to 'go nuclear', either for defence purposes or in an attempt to develop a viable deterrent force. Both paths offer serious technical problems and for neither is it clear that a force can be developed which will appear credible as a deterrent or a defence against the United States or the Soviet Union – unless it is done in an all-European context. Since the development of either type of nuclear force carries serious political risks, e.g. greatly increased military budgets and worsened relations with neighbouring nations, it is doubtful whether there will be any rapid decisions to acquire nuclear forces. Over a longer time period, however, and in context of a continuing arms race between the United States and the Soviet Union, the pressures towards a political restructuring of European alliances and towards the development of some kind of independent nuclear forces will probably be irresistible. Put in other terms, it appears almost inevitable that a continued Soviet–American arms race, with its clear implication that the two superpowers are giving an overriding priority to their own security, will cause a drastic weakening of the current NATO and Warsaw Pact alliance structure. Where and by what means the European nations would find their security is much less clear. One can hope that it would first be sought in a strengthened and envigorated United Nations rather than in a purely European context but this too may be unreal in light of a continued Soviet–American arms race.

D. CARLTON

# 2: Anti-Ballistic Missile Deployment and the Doctrine of Limited Strategic Nuclear War

In recent years it has become fashionable to refer to the 'nuclear stalemate', to the 'balance of terror' and even to the 'nuclear age' as if our brief experience of these phenomena enabled us to face the future with a well-founded confidence that the possibility of major wars between the superpowers can now quite definitely be ruled out. Yet in the late 1950s important sections of public opinion in the Western world reflected an entirely contrary expectation. The news of the first Soviet Sputnik engendered in the United States a startling concentration on pessimistic preparations, symbolized by the construction of fall-out shelters; while in Great Britain and Scandinavia campaigns against the manufacture and possession of nuclear weapons achieved surprising momentum. However, once it had become apparent that the prophets of *imminent* catastrophe had been mistaken, there occurred an equally marked reaction in the direction of complacency; and the outcome of the Cuban confrontation of 1962 has tended to confirm the opinion that a major trial of strength will not occur in the foreseeable future. This might therefore be an appropriate juncture to consider whether such optimism can seriously be challenged.

The principal criticism that may be advanced against the optimists' position centres upon their assumption that we have been living in the Nuclear Age for two decades already and therefore have sufficient evidence to draw some empirically justified conclusions about our present environment. Against this it may be argued that the Soviets have had a serious potential to inflict unacceptable damage in a first strike on American cities for only a decade and that a comparable Soviet second strike capacity is of still more recent vintage. Our experience so far therefore is of a world in which one superpower has had a commanding lead in the nuclear field partially offset by its rival possessing advantages on the conventional side. The so-called military balance has in fact been to a great extent advantageous to the West. That this has not been adequately demonstrated by successful acts of conquest probably owes more to American self-restraint than to any Soviet deterrent. If the military roles had been reversed it is to say the least not a self-evident truth that similar restraint would have been shown by the Soviet Union. However that may be, it is apparent that we have as yet no considerable experience of the nuclear age proper – that is when the two superpowers each possess invulnerable second-strike capabilities. The first two decades after Hiroshima no doubt involved considerable perils and it would be foolhardy to claim that a major conflict was never a possibility. Yet the important thing now is not so much to congratulate ourselves on weathering the storm for twenty-three years – though we have clearly done so – but to recognize the unique character of these years of prologue. We are now embarking upon a voyage into an uncharted sea, where our previous experience will be of very limited relevance.

In the changed circumstances the quintessential problem we face concerns the utility of the strategic nuclear threat. More precisely, can a superpower any longer rely on a declaratory policy of 'massive retaliation' to deter anything other than an all-out attack on all its cities by its rivals? There can be in fact no rational expectation that such a policy would work successfully for very long. That is not to say that we must expect a deliberate and calculated attempt by any Soviet or American leader to explode the myth of 'massive retaliation' as a deterrent to all major acts of aggression. Rather should we anticipate a steady erosion of faith even in the declaratory policy and the growth of a tacit recognition on both sides that limited war has become an acceptable method of settling outstanding quarrels, always stopping short of any attempt by this method to conquer the homeland of a principal rival.

What form, then, may we expect such limited wars to take?

If East and West were in a situation of approximate parity for undertaking trials of strength at the conventional level, we might look for a tacit acceptance of a permanent ban on the use of nuclear weapons. But this parity plainly does not exist. The Western Powers, being subjected to electoral pressures, would undoubtedly have great difficulty in introducing conscription on the scale needed to match potential Soviet and Chinese efforts. And even if this were achieved, real parity would not exist because of the geographical advantages possessed by the Communist camp. The Western Powers are obliged to defend innumerable outposts on the periphery of the rival landmass, whereas only Cuba is a comparable Achilles Heel for the rival alliance. It seems apparent therefore that to redress the imbalance Washington will have to be prepared in certain circumstances to have recourse to some form of limited nuclear response.

In deciding upon the form of nuclear power to employ, the Americans would have to decide whether the aim was to compensate for conventional inferiority in the theatre under attack, or whether the objective was to force their rivals to return to the *status quo ante* following a token demonstration of resolve in a global context. In the face of unambiguous and determined aggression the choice would probably lie between the introduction of so-called tactical nuclear weapons in the relevant area or the employment of a teaching strike on an enemy city far-removed from the scene of the conventional struggle. Of the two the latter would seem to be the most rational response for a defensive power suffering from conventional inferiority. The use of tactical nuclear weapons would probably lead to retaliation in kind with the likely outcome being a stalemate but not the restoration of the strict *status quo ante*, since the territory being defended would have been reduced to a depopulated wasteland – an outcome from a strategic viewpoint only marginally better than its loss intact to the enemy. Moreover, the use of tactical nuclear arms in the field would probably lead to an unprecedented growth of panic and insubordination at all levels of the armed forces of both sides, with the attendant risk of rational central direction of the conflict being lost.* The only viable alternative in some circumstances might well be controlled counter-city teaching strikes. Needless to say, the city or cities chosen for such destruction would not be likely to include Washington or Moscow.

The cardinal advantage of this particular strategic posture, especially when compared with other nuclear strategies, lies in its credibility. The employment of teaching strikes in any particular conflict would mean that the outcome would not depend on military capacity either on the conventional level where a decisive disparity may exist, or on the level of total strategic nuclear capability. Instead the emphasis would be almost exclusively on the rational display of willpower and resolve. Certainly there would appear to be a good chance that the strategy of exchanging teaching strikes would end not in Armageddon but in the restoration of the *status quo ante*. Of course those who decided to make the initial destabilizing move would need to recognize they were testing their opponents' resolve and be prepared ultimately to withdraw if this proved to be adequate. What would constitute an 'adequate' display of resolve is impossible to know in advance, but it may be that nothing less than the elimination of a sizeable city would carry much conviction.† Furthermore, what sufficed in the first such encounter would probably prove inadequate on a subsequent occasion. In addition there might be considerable domestic complications but, given adequate pre-planning, these aught not to be of such an order as to seriously inhibit the decision-makers. It is a matter for speculation whether the closed societies of the Soviet bloc or the libertarian states in the West would be better equipped to handle the volume of public protest which would accompany such encounters.

* The same drawback applies to the erstwhile McNamara notion of engaging in limited counter-force strikes.

† The Cuban Affair of 1962 showed distinct promise of developing into a serious conflict of resolve but Khrushchev tantalizingly backed down before the encounter was properly under way. Perhaps a disparity in strategic nuclear strength was a discouraging factor. The Chinese polemicists have also pondered on this problem but cannot quite decide whether Khrushchev was a 'criminal adventurist' or a 'criminal capitulationist' or even both in turn.

A more serious problem arises if one set of decision-makers will not play according to the rules. Should we in fact take seriously the Soviet declaratory insistence that if one nuclear weapon is ever used, 'the balloon goes up'? Fears on this score tend to recede rapidly once it is realized that the Soviets, with their advantage at the conventional level and their impending approach to effective parity in second-strike strategic nuclear potential, have every reason to wish that limited strategic nuclear war should be ruled out as a serious option.* We may console ourselves with the knowledge that they have certainly digested the sophisticated American literature on this subject and that they are unlikely to have found a rational refutation. In practice they are therefore no more likely than the Americans to act irrationally and initiate Armageddon.

The same unfortunately cannot be said for any smaller power possessing the capacity to deliver strategic nuclear weapons on Washington or Moscow. For example, a spasm response remains the only credible nuclear threat *vis-à-vis* the Soviets open to Great Britain, currently the only country in a position to take such action. Geographical and strategic nuclear disparity make it inconceivable that the British alone could seriously engage in a protracted duel of controlled teaching strikes with the Soviets.† Of course precisely because a spasm response is for Great Britain in a supreme national emergency involving the superpowers the only real alternative to surrender, the threat of the former carries a degree of credibility to which the Americans and the Soviets cannot now aspire. It might even be said that in this way the British nuclear capacity provides an additional deterrent for the West as a whole and one which incidentally the Soviets cannot match, since they appear to

* In this respect the Soviet attitude is probably of exactly the same character as Western threats to launch all-out nuclear war in the event of a major conventional attack, namely an increasingly unconvincing bluff.

† The British Polaris fleet about to come into service consists of a mere four submarines of which one or two will always be out of commission for servicing. Thus the British could have little hope of enduring more than three rounds of limited counter-city exchanges, since each use of a single missile would reveal the location of a submarine, which would then be an easy target for Soviet attack submarines.

have no ally which is simultaneously both independent and reliable. Perhaps after all President Kennedy knew what he was about at Nassau!

Yet the deliberate retention of a relatively credible though ultimately irrational deterrent in the West cannot be deemed wise. There must always remain a finite chance that even British leaders will act irresponsibly and, more important, it represents a partial recognition of the value of irrational threats when all the West's efforts should be being channelled in the opposite direction. By renouncing irrational options for themselves the NATO powers could, for example, take a decisive step towards discouraging experiments in simulated insanity by the Soviets. The Americans in such circumstances could convincingly argue that they would take no notice of an opponent's possible insanity as distinct from legitimate resolve. It would then clearly be for subordinates to remove a genuinely demented leader by whatever means were necessary. Yet the Soviets are unlikely to be convinced of the futility of reliance on irrational and 'insane' threats unless the Western Allies themselves make a clear break with them. It is therefore of prime importance, if this view is accepted, that the Nassau agreement be suitably modified and the Wilson government's declaratory policy translated into physical reality.*

Many liberals and progressives both in Great Britain and elsewhere would of course favour the denegotiation of the Nassau Agreement. They might not do so specifically for the reasons outlined above. But they would for the most part accept the undesirability of minor nuclear powers being able to trigger off Armageddon by delivering on a superpower's capital a spasm attack, which the recipient might find indistinguishable at first from a first-strike attack by the rival superpower. Opposition to such a catalytic use of nuclear weapons was, for example, widely voiced in Great Britain by Labour and Campaign for Nuclear Disarmament spokesmen in

* The case for British retention of nuclear bombs and obsolescent V-bombers for possible use in a 'Second League' of minor nuclear powers is of a different order. A British force of that character would represent no very important factor in a Soviet–American confrontation.

the last years of Conservative Government.* The same progressive elements would also doubtless welcome a unilateral withdrawal from the nuclear arms race by France and China. Again, and for similar reasons, they would tend in addition to give strong support to the movement for a Non-proliferation Treaty.

In all these respects of course they have a powerful ally in President Johnson. He apparently sent George Ball to London in 1967 to make unofficial inquiries as to whether the British Government would be willing to modify the Nassau Agreement; he has indicated his disapproval of Gaullist policies in general; and he has proved to be a fervent supporter of a Non-proliferation Treaty. Yet there is one element in Johnson's policy of attempting to confine the power to launch Armageddon to the Soviet Union and the United States which has not received the general endorsement of progressive and liberal forces in the Western world. This is of course his decision to deploy a limited ABM system directed ostensibly against China but in reality against *N*th powers in general. We must now examine the reasons for this divergence of approach between much of progressive opinion and the Government of the United States.

Some of the critics of President Johnson's ABM deployment base their case on a fear that superpower ABM race is occurring or will develop and that this will certainly put an end to the allegedly existing *détente* and may quite possibly lead to a major nuclear war. Others have criticized the expense involved; some have doubted the technical effectiveness of ABMs even against minor nuclear powers; and yet others have denied that any *N*th power would in any circumstances launch a suicidal 'spasm' strike against a superpower. All these arguments have some weight and deserve close analysis. But before attempt-

* For example in January 1964 Denis Healey wrote as follows: '. . . so long as the Prime Minister [Sir Alec Douglas-Home] insists that you cannot trust the Americans to come to your help in a crisis and that therefore you must have atomic weapons in order to trigger off the American Strategic Air Command against the will of the American Government, he is strengthening and accelerating that very trend in the United States to reduce American's liabilities in Europe which is the excuse for his position.'[1]

ing such an exercise it is perhaps permissible to suggest that for many critics of ABM deployment these arguments are a mere rationalization of a deeper and fundamentally emotional response. The fact is that many progressives accept as an article of faith that *any* escalation in the so-called Arms Race is undesirable and wicked. This conviction is based on a number of confusions and on a misreading of history. It is not true, for example, that armaments or arms races in themselves inevitably cause wars. They may or may not be a secondary causal factor depending on the circumstances. But from past experience it may be claimed that the *absence* of certain armaments is at least as likely as their existence to create a situation in which war is made more probable. In general it seems evident that a power structure based on an uncomplicated bi-polar military equilibrium is likely to be more safe than one in which, as in 1938–9, there is a fundamental lack of balance or one in which, as in 1914, there is a complicated power structure containing possiblities for catalytic action by powers of the second rank. If it is conceded that the present bi-polar system is preferable to that of 1914, then it is surely necessary that, in order to maintain it, the ABM escalation should now take place. Some progressives would of course concede that the effectiveness of a deterrent system based on an equilibrium is not necessarily related to the nature of the armaments involved, that is to say they do not believe it to be a self-evident truth that a balance of terror based on sophisticated offensive and defensive rocketry is *ipso facto* less stable than one based on old-fashioned manned bombers. But many would nevertheless argue against a *pari passu* escalation, as in the present case of superpower ABM deployment, on the grounds that disarmament becomes progressively more difficult to achieve. This superficially attractive view is again quite fallacious. It owes much to the belief that the achievement of a disarmed and peaceful world depends on slowing down the Arms Race, halting it, then reversing it, and finally, like a motor car, driving it into the garage of GCD and permanent peace. The unwelcome fact is that our regrettable failure to achieve GCD has nothing to do with the speed or level of the so-called Arms Race. This failure stems instead from the fundamental inability or unwillingness of the existing sovereign states to agree on terms for ending the international anarchy. In fact men have lived in a condition of international anarchy from the beginning of recorded history and have never come near to replacing it by World Government. In this sad record fluctuations in the speed of the Arms Race have made absolutely no difference. It is thus clear that many progressive thinkers in their obsession with the Arms Race are attacking the symptoms and not the cause of the continued existence of wars and the threat of wars.*

Having dealt with the irrational basis for much of the opposition to ABM deployment, we may return to the more substantial arguments as previously outlined. First there is the contention that the expense involved is disproportionate to the benefits (if any) that would accrue from the deployment of ABM systems in the Soviet Union and the United States. Against this view it may be retorted that any expenditure that even marginally adds to the chances of the survival of the human species is almost self-evidently worth while; and that if the superpowers really need to make savings, they could forthwith terminate their respective wasteful space programmes.

A second argument against ABM deployment relates to the alleged incredibility of any nuclear threat to the superpowers by China and other second rank states. It is indeed unlikely in the extreme that even China would launch a *first-strike* nuclear attack on the United States or the Soviet Union. But a Chinese *second-strike* retaliation against a nuclear attack by a superpower is surely not so inconceivable. Many commentators would argue of course that it would be the height of immorality

* It should not be concluded from the foregoing argument that it follows that one should be opposed to agreements on arms limitation and/or reduction. It would follow, however, that one would see such agreements have nothing to do with and do not bring nearer such millennial objectives as World Government or the Reign of Perpetual Peace. Once this is recognized it is of course possible to judge each limited agreement or proposal on its merits as a contribution to the preservation of the existing uneasy peace. Many such measures would help to that end, but others, like a superpower ban on ABM deployment at the present time, may have a contrary effect. In every case the balance of risks needs to be judiciously weighed.

for the superpowers to deploy ABMs so as to be able with impunity or near-impunity to threaten a first strike nuclear attack on China.* But is such a justification really so worthless? How, in the absence of such a threat by the United States, could the United Nations forces have prevented South Korea from being overrun in the early 1950s? Similarly, if in the present circumstances the United States cancelled her ABM plans, how long would hopes of Asian stability survive? It seems clear that as soon as China is able to threaten American cities with unacceptable damage the tacit American nuclear guarantee of non-Communist Asia will become very much less credible than it is now. Two consequences might follow. First, it is conceivable that the Chinese, using nuclear blackmail, could sweep across much of Asia without any serious challenge from the United States. Secondly, and more probably, such Asian countries as India, Japan and Indonesia, fearing the first possibility and notwithstanding the Non-proliferation Treaty, might decide to make their own nuclear weapons. If this occurred, it would be a repetition both in form and motivation of the Gaullist development in Europe, albeit in a far more unstable theatre.

A third argument against ABM deployment is that it would have deleterious effects on the relations between the superpower and might even lead to a global war. Here it must be conceded that there is indeed some slight risk that one or other of the superpowers might achieve a breakthrough and outstrip the other in invulnerability or, even worse, might seem about to do so. This would certainly increase the chances of a preventive strike being launched by the power that feared eventual military eclipse; and a belief that such a strike was imminent might equally lead the other power to launch a pre-emptive attack. But none of this seems likely, since, if McNamara's

* On the moral question it is worthy of note that Michael Stewart, the British Foreign Secretary, has written: 'It can hardly be contended that the West is bound by some rule of chivalry to keep war on the conventional level. The guilt of war lies on the aggressor, and its wickedness lies in the fact that it is an appeal to physical force rather than to reason and justice: the power that makes this appeal cannot then claim that the conflict shall be conducted only with those weapons that suit it best.'[2]

testimony is to be believed, offensive preparations by a superpower will for the foreseeable future easily cancel out the more expensive defensive preparations of its rival. There seems therefore to be no good reason for anticipating that ABM deployment will bring about any significant deterioration in relations between the superpowers; and such risk as there is appears to be well worth running when compared with the far greater risks involved in allowing the present relatively stable bi-polar military system to be superseded by a polycentric version.*

Finally, we must consider the argument of those who deny that ABM deployment has any considerable military value against the offensive capabilities of even a primitive nuclear power like China. In the last resort this contention can neither be substantiated nor dismissed with any degree of certainty, owing to the existence of conflicting evidence on the technical aspects.† But it is possible to guess that ABM systems would at very least give the superpowers marginally more bargaining leverage *vis-à-vis* the minor nuclear powers. For example, American nuclear blackmail of China is more likely to be effective in a situation where neither set of leaders knows with certainty whether China could or could not deliver even one nuclear weapon on an American city than in a situation where it was known beyond doubt that she could inflict an unaccep-

* The best hope of minimizing the risk of a destabilization of the central balance would probably lie in a formal or tacit agreement between the superpowers on the following four-point programme:

(i) A 'thin' ABM defence of major cities so as not to devalue the rival superpowers' second-strike counter-value threat;

(ii) A 'thick' ABM defence of land-based ICBM bases so that the recent MIRV development should not make a first-strike counter-force attack more attractive;

(iii) A rapid shift of resources to increasing the number of rocket-firing submarines so as to strengthen the invulnerable second-strike counter-value capability of the superpowers;

(iv) A moratorium on the construction of attack submarines so as not to threaten the reliability of the rocket-firing submarines.

† It is, however, worth noting that pessimism regarding defensive capabilities is nothing new. In the 1930s many commentators shared Stanley Baldwin's expectation that 'the bomber would always get through'. Many believed that a second world war was unthinkable, since all major cities would be totally destroyed in the first few days of the conflict.

table degree of damage. In addition it must be stressed that to concede that the ABM systems of the superpowers may not be 100 per cent effective against *all* minor nuclear powers for the rest of the century is not to concede that such systems are of no value. They will surely have been worth while if they delay by only a decade the day when the two superpowers become vulnerable to other nuclear powers. They will also surely be of value even if after a decade they are only able to limit the number of such powers to one or two rather than to three or more. Finally, it is possible here to justify ABM deployment in the context of the doctrine of strategic nuclear war, as previously described.* Without such a deployment Washington and Moscow will soon become vulnerable to catalytic and spasm attacks from at least Great Britain and China – and no doubt from other states as well in due time. If, for example, in the 1980s the German Federal Republic were able to deliver nuclear bombs on Moscow, a Soviet-American duel over Berlin might quickly escalate into all-out nuclear war without either superpower desiring it. With ABM deployment, however, it should be possible to safeguard at least the capital cities or other key decision-making centres of the superpowers against such attacks into the indefinite future. Hence it ought to be possible for Soviet and American decision-makers, safely ensconced in their capitals or their alternative retreats, to exchange counter-city teaching strikes without having to worry about threats by minor nuclear powers to precipitate Armageddon by blowing them up.†

What then are the prospects for war and peace in the nuclear age which is currently dawning? We have concluded that there is no valid justification for facile optimism and complacency in view of the considerable possibility of a major Soviet-American test of resolve eventually occurring. On the other hand, it is also clear that if the superpowers can by ABM deployment

* For a fuller description of the doctine and its implications see Klauss Knorr and Thornton Read (eds.), *Limited Strategic War* (Princeton, New Jersey and London, 1962).

† It is of course central to the doctrine of limited strategic nuclear war that the decision-makers should not be subject to vexatious panic-inducing threats to their personal safety.

remain largely invulnerable to all other nuclear forces, the chances of all-out strategic nuclear warfare ought rapidly to recede as the case for a more rational response to a major challenge gains acceptance in Moscow as well as in Washington. Paradoxically, this is likely to increase the chances of lesser wars taking place. But this is a price we should be prepared to pay if we wish to lessen the chance of the total destruction of our species, which is implicit in doctrines of 'massive retaliation'.

Of course no particular war is inevitable until it has happened just as no particular road accident is predictable in advance. Yet just as we may feel certain that road accidents *per se* will not become impossible for as long as motor cars are allowed to compete with each other on communal roads, so we may expect occasional military encounters for as long as a plurality of totally sovereign states share a single planet. In the long run there is no reason to suppose that the invention of doomsday weaponry will decisively undermine this simple and ancient truth.

*References*

1 D. Healey. *A Labour Britain and the World*, Fabian Tract 352, London, 1964, pp. 13–14.
2 M. Stewart. *Policy and Weapons in the Nuclear Age*, Fabian Tract 296, London, 1955, p. 6.

M. LEITENBERG

# 3: Anti-Ballistic Missile Deployment and China[1]

> We can begin by discussing China's possible reactions. It is doubtful that anyone seriously believes that the ABM, as currently described, is aimed at China. However, since public justifications have insisted that the essential aim of the ABM system is deterrence or defense against Chinese threats, the argument must be examined. From 'The ABM, Proliferation and International Stability', R. L. Rothstein, *Foreign Affairs*, April 1968.

The entire question of United States ABM deployment against, or in relation to China depends, or should depend, on assumptions regarding the aggressiveness of China. This disregards for the moment the planned disposition of the Sentinel system to protect American Minuteman ICBM sites. Recent newspaper reports have indicated that the United States Department of State has succeeded in temporarily postponing the formal decision to this effect; however, the Sentinel system has so far been consistently publicly described by official sources as providing for ABM batteries for the protection of ABM radar and ICBM sites. For this and other reasons very few, if any, experts believed the repeated United States Government insistence that the Sentinel ABM system was intended as protection against Chinese attacks. Private conversations with Department of Defense officials in Washington in the last week of December 1967 ascertained that DOD planning had already designated the twenty-five cities that would be supplied with ABM defences under a 'light' system and had provisionally budgeted for the procurement of missiles for this requirement in the budgets to be presented in 1969–71. Testimony by

McNamara, Foster and General Wheeler in February 1968 before the Senate Armed Services Commiittee and released in April, for the first time uses the phrasing that the Sentinel system is 'consistent with the first phase of a deployment against the Soviet Union'. The private conversations are, however, presently unverifiable.

Sources for and the registration of disbelief of the officially pronounced motivation have come from spokesmen not customarily given to such dissociation from official positions. It is necessary to indicate their degree and nature at the outset, lest the chapter be criticized for naïveté.

*Senator Kuchel:* In the recommendation of the Joint Chiefs of Staff was any consideration given to the problem of Red China?

*General Wheeler:* We took a look at it; yes, sir. However, we do not believe that we should deploy at this point in time an anti-ballistic missile system purely to defend against the Red Chinese threat.[2]

Washington. Johnson Administration's decision last fall to recommend a limited Sentinel anti-ballistic missile system for area defense of the U.S. against a possible future Communist Chinese threat took defense planners by surprise. Previous Nike-X deployment studies had emphasized primarily the protection of Minuteman sites and other strategic military facilities against a Soviet intercontinental ballistic missile (ICBM) threat.

Recently released congressional testimony appears to support the view that the Sentinel decision announced Sept. 18 was at least partially intended to deprive Administration critics of a possible election issue comparable to the 'Missile Gap' of the 1960 campaign.

......

In testimony before the House Armed Services Committee, Lt. Gen. Alfred D. Starbird, manager of the Army's Sentinel System project, acknowledged that 'some changes to the deployment model resulted from studies undertaken after the deployment decision was announced by the secretary of defense. A complete recosting of the program was necessary because of these changes. This recosting is under way but has not been completed.'[3]

Because he couldn't persuade the Russians to consider limitations on missile defenses, the President has now ordered the building of a 'thin' defensive system to protect us from the Chinese. The logic of the President's decision seems mighty tortured.[4]

Mr McNamara's San Francisco speech in which he announced the American decision to deploy a light anti-ballistic missile system as a defence against China is a very important statement of America's nuclear policy. The force and clarity of much of Mr McNamara's strategic thinking, however, has obviously had to be blunted by the less precise demands of American domestic politics, and it is finally as a domestic political gesture that one is forced to regard the ABM decision.[5]

If a US President authorizes a $5 billion Sentinel system to protect himself from Republican charges of failing to insure the nation's security, then one might just as well junk all the elaborate systems of defense analysis that we possess.[6]

To be sure, the Administration asserts that our ABM system is being deployed to counter the possibility of a light Chinese attack in the coming years. The Chinese are testing thermonuclear devices and medium-range missiles; within the next decade they could achieve a modest ICBM force. In the same budget testimony, Secretary McNamara argued that the proposed ABM system might hold fatalities from potential Chinese attacks of varying plausible sizes in the next five to eight years to below a million; in the absence of such a system the figure could range from seven to fifteen million. But why deterrence on which we have relied successfully for many years to restrain the powerful forces of the Soviets, is not expected to work against the small forces for the poorer and weaker Chinese is totally unclear. For some time into the future, at least as far as our present projections and calculations go, we shall have a credible first-strike capacity against the Chinese. In such circumstances, it is difficult to understand on what calculation the Chinese would throw away the only effective restraint on our nuclear force, the strong moral and political inhibitions we have against striking first, by launching an attack.

Only the proposition that there is a special danger of aggressive irrational behaviour by some future Chinese Government would make sense of the official argument. Yet there is nothing in the recent record of the Chinese Government in its international relations that provides the basis for such an expectation. Accordingly, it is hard to take the Administration's rationale for deploying a thin ABM defense at face value, especially in the light of its own evaluation of the utility of not only the proposed defense system, but more elaborate, extensive and powerful ones. The decision is much more easily understood as the result of long-standing pressures within the military establishment and the Congress for an antimissile defense. When these were reinforced in recent years by the evidence of initial Soviet deployments, the demand for a 'response' finally became irresistible.[7]

Why then discuss the United States ABM decision on the merits of its anti-Chinese capability at all, instead of devoting one's energy and attention to its domestic political or anti-Soviet aspects? It seems that at some point one must attempt to evaluate the validity of the arguments that were officially presented. What permitted the United States administration to presume that it could offer this explanation of its decisions to the American, Chinese and Soviet publics; to the Chinese and Soviet administrations, and to the world in general? It will be necessary to look at some of the American perceptions of Chinese foreign policy since 1950.

The entire construct of 'Chinese irrationality' which for so long has been the twin of China's aggressiveness in public statements by American officials, warning of what America and the world had to fear and face in China's potential behaviour, is a rather transparent piece of semantic nonsense. In fact we may have just seen its official demise as a public justification. For years professional sinologists have insisted that China, although verbally hostile and making extraordinary use of violent language, had shown a record of actions (or non-actions) showing a real appreciation of American power and avoiding external adventures that would risk a direct confrontation with American forces. In opposition the United States administration presented the view of China as rampagingly aggressive. This view was maintained through the debates about ABM, though McNamara made clear the American intention or willingness to demolish China for the slightest 'irrational' nuclear attack on her part against the United States. In November 1967 Deputy Secretary of Defense Paul H. Nitze's testimony contains the following remarks:

> The question naturally arises as to why we do not rely on our capability for assured second strike destruction to deter the Chinese People's Republic (CPR) as we do rely upon it to deter the Soviet Union. The primary difference between the Soviet and the CPR case is that it is feasible to provide a damage denial ABM against the CPR but it is not feasible to do so against the Soviet Union. If we could assure effective defense of our population against an attack such as is possible with Soviet capabilities we could recommend that we do so. But at the present state of technology, we cannot. Against the CPR, we can. . . . This characteristic of a small, vulnerable system could create pressures toward reckless behaviour, even in a people not by nature reckless. That pressure could make people who are not by nature reckless undertake more reckless acts . . . our deterrent power against China would probably be effective, but you could not be sure that this other effect might not make even a cautious people more reckless.[8]

China is thus officially and quickly transformed into a 'people who are not by nature reckless' and 'a cautious people'. The only previous official presentation of this notion was given by William Bundy in April 1966 among the more typical remarks from which one would conclude that China was everywhere pressing on the United States. Buried in the paper is one sentence: 'and it can certainly be argued that they are tactically cautious in pursuing these ambitions.' This is immediately dropped and the paper returns to a discussion of 'the extent of their ambitions'. The extent of 'pursuing' can be objectively assessed, the 'ambitions' cannot, but Mr Bundy is thus allowed to dispense with any relevance to reality in the remaining portions of the paper.[9]

If one took seriously the United States ABM as an anti-

Chinese weapon it would do much to undercut the concept of deterrence as a strategic justification, despite Mr Nitze's disclaimers. If one assumes the United States ABM to be anti-Soviet the same concept seems to have been deserted to some degree as well. What may be more relevant in Mr Nitze's statement is what one can extract from the word 'feasible'. The implication is that one wants to have absolute confidence in the continuance as long as is possible of the present United States force posture against the Chinese. If a conflict should break out at any time before China had a more advanced or ABM penetrable nuclear delivery system, until then one would want to be able to act militarily *vis-à-vis* China without any fear of Chinese damage to the continental United States of America. After that time problems of ABM versus Chinese delivery systems would be the same as those of ABM versus Soviet delivery systems, and perhaps the question of deterrence might return to being relevant. Support for this view comes from one of the few available sources in which the strategic arguments eventually offered for the Sentinel decision first appeared:

> A profound and radical shift in military thinking on ballistic missile defense has been taking place in the Pentagon over the past year. The change arises in part from a new view of the future ballistic threats which may confront the United States as a result of the Red Chinese nuclear detonations, and in part from a new appraisal of the rate of growth of ICBM defensive capabilities relative to the offense. Detonation by the Chinese of two 'nominal' nuclear devices during the past 12 months has raised anew the specter of nuclear weapons proliferation . . . in the narrower view of the Pentagon the principal danger of proliferation is that it would tend to reduce the scope of United States overseas initiatives and therefore the power of the United States. Consequently, there has occurred an abrupt reappraisal of our effort to devise a ballistic missile defense, instead of viewing it in terms of its performance against much reduced threats variously termed small, moderate or even 'primitive'. It has been discovered that a relatively modest deployment, covering perhaps 40 cities at a cost of $6–$8 billion, could save many millions of American lives which would otherwise be lost in even a small scale attack.[10]

This line of argument might imply that China could even argue that the United States expected or intended armed conflict with her before some estimated date. At least we seem to want to retain the option for that possibility as long as possible.

Several remarks about the concepts of deterrence and stability, which enter so often into these discussions, seem necessary. It was in 1961 that Schelling and Halperin suggested that the United States should convince the Soviet Union of the need for hardened missile sites and 'wait' for them to be built.[11] In 1967 Dr Fink (former Assistant Director, Department of Defense, Research and Engineering) commented that everyone agrees that Polaris is stabilizing, but that no one suggests stopping the wild rush to improve anti-submarine warfare systems, and everyone is rushing to improve them. No one has ever suggested doing away with hunter-killer submarines. No one has ever suggested giving the Soviet Union (or the Chinese) several Polaris submarines. All of these suggestions are entirely compatible with deterrence on logical grounds, yet few American strategic observers would support or even consider them. Exactly what is the meaning of the deterrence concept then? To what degree does it or does it not depend on logic? To what degree does it incorporate expectations of use? To what degree are the actions carried out in its name really aimed at fixing a *status quo*, or to what degree are they destabilizing? What will a breakthrough in anti-submarine warfare for either side do to the Polaris situation? Is 'deterrence' deterrence if it is not bilateral?

The lesson of the Cuban missile crisis was that the United States was prepared to push the buttons and fire her strategic missiles. We have for years now, at least in public diplomacy, claimed the function of mutual deterrence. There is always the problem of the balance between judgment and 'insurance', and no one suggests the waste of 10,000,000 or 40,000,000 lives as payment for human error. The answer must be that China would not be willing to trade a major portion of her population to sting the United States, and that in the end other policies and other attitudes would be better insurance (certainty, longer lasting, less costly) than technical hardware. A development of such mutual confidence might prevent any loss of life, including those of conventional engagements, and return us to expectations of Chinese, or another nation's, behaviour on bases of 'rationality', rather than irrationality and 'paranoia'. The logic of irrational expectations in fact predicts that Chinese attack would come no matter what insurance we prepared or what befell them in return. Important questions are:

(*a*) What can one nation posit concerning a second nation's behaviour if one looks for the sources in irrational or 'pathological' approaches?

(*b*) Does our Department of Defense and the administration really believe the requirements of its own proclaimed logic concerning such Chinese pathological motivations? If Mr Nitze's change of emphasis makes these questions now seem less necessary, one has only to recall how recently the previous phraseology was current, that one in fact does not know what assumptions are behind present American strategic policy towards China, nor what they really were before, and that if such assumptions were similar to the publicly professed ones, the American foreign policy and decision-making apparatus is not known for rapid accommodations to new themes in belief or practice.

In looking at the question of China, the United States and the ABM, it is relevant to consider, among other things: the relevance of China's conflicts and other conflicts elsewhere in the world in making a judgment of threat; American presentation or perception of these conflicts and threats; China's threat to the United States; China's threat to her neighbours; China's military capabilities outside her borders; the relevance of these to the decision concerning an ABM; and the effects of the decision.

China is no threat at all to the United States. She is militarily weak as major power capabilities go, judged by nearly every military criterion available except population number and, perhaps, will. Given the strategic retaliatory capability of the United States, this situation is not changed by China's development of nuclear weapons. As far as her neighbours are concerned one cannot tell whether her intentions would ever again involve 'aggression' or military pressure.

There have been recent public reassessments of the conventional force balance in the European theatre.[12, 13] Apparently, something similar may even be occurring concerning China and her capabilities beyond her borders:

> While China would probably prefer to expand her influence through insurgencies, she might turn to a direct attack on Korea, Taiwan, India or Southeast Asia if an insurgency were failing and she was willing to risk overt aggression.
>
> At first glance the size of the Asian Communist forces (CPR; North Korea, North Vietnam – 3 million men) suggests that it would be nearly impossible to stop such an invasion.
>
> . . . . . .
>
> In fact, however, the Red Chinese have only a limited ability to attack beyond their borders. First, there are great barriers between China and her neighbors; the Himalayas, the jungles of Southeast Asia, and the Formosa Straits. Second, because the Chinese soldier is not nearly as well-equipped and supplied as his American counterpart, he is far less effective in conventional combat. Some indices of the relative firepower and mobility of Chinese and U.S. soldiers are shown in the next table. The average U.S. soldier has three times the fire-power, five times the motor transport and twenty times the equipment of a Chinese soldier.
>
> Finally, the Asian Communists have limited offensive air ability. The MIG 16s, 17s and 19s, comprising 85 per cent of the Chinese Air Force, cannot attack targets much beyond the borders of China because of their limited range and the location of Chinese airfields.
>
> When one examines the invasion threat on a theater-by-theater basis it is clear that despite the huge Chinese Army, existing U.S. and local forces provide both a strong deterrent and the ability to defend important areas [a theater-by-theatre survey followed].[14]

Coffey has suggested that advantages that would result from deployment of ABM system *vis-à-vis* China would be: the bolstering of the American deterrent in the eyes of our Asian Allies; could help preserve American freedom of action *vis-à-vis* China; and materially reduce the casualties from a Chinese ICBM and SLBM attack.

The third condition would only arise from the second. American freedom of action is now total *vis-à-vis* China. It is presently limited only by international opinion, and mutual defence clauses still extant in Chinese treaties with the Soviet Union,[15] and it is doubtful if it will be limited by much else before China achieves, if she does, a full-fledged delivery system. The 'credibility' of the American deterrent needs no bolstering in the eyes of our Asian allies. There is a point at which deterrent cannot be bolstered; it can simply get no bigger. Somewhere along the development of a Chinese nuclear weapon the theory of the deterrent function of the deterrent force seems to have been implicitly downgraded. The United States does not need protection against minor, 'smaller scale' Chinese attack. The protection exists in the present American orientation to the Chinese. Such attack would be suicidal on China's part, and it cannot possibly be imagined that they do not perceive this. If our deterrent is not credible now, it never will be, and fantasies of deficiencies and chinks will always be present, and with time and proliferation ever more so. 'Total' (as opposed to 'smaller scale') attack by the Chinese would of course incur the same American response but would some day probably be more involved with and dependent upon our posture and attitude towards China than on the magnitude of the deterrent or its ABM enhancement. And on this last point, we may be bringing about, perhaps to our utmost, the very thing we claim to fear.

Discussions of ABM and China often infer the potential utility of Chinese nuclear attack against Asian allies of the United States such as Japan, the Philippines or India, or against American installations in the Pacific by MRBM. A United States based ABM system would of course be of no direct utility in this regard; it is said to be only added indication of our willingness to undertake nuclear retaliation against China for such 'hostage' attacks. Such cautious and expert commentators as Coffey and Hsieh[1, 15] find it realistic to consider and discuss such a utility for Chinese nuclear weapons. There enter here the arguments of credibility of threatened all-out nuclear attack as a response for lesser provocations and other arguments of strategic doctrine; however, this entire 'hostage' discussion is puzzling. Given the continuing assumption that the mood in the United States would predict a nuclear attack on China of some degree, perhaps massive, in retaliation for her use of such weapons against an American base or ally, it seems ridiculous to presume that China would draw on to herself such an attack, which she is avoiding, and which she would earn by an attack on the continental United States itself, by an attack on Camhranh Bay or Clark Field. Chinese nuclear 'threats' to Camhranh Bay are not forthcoming and, if they were, would not 'cause American decision-makers to hesitate', at least not in the near future. Levels and duration of the response of the United States to the Tonkin Bay and Pleiku incidents support the contention that this response would be strategic, and not tactical. American public opinion, if at all relevant, would not object, nor would there be much concern about 'history'. Again the evidence of our willingness to use a nuclear response to a non-nuclear attack at Quemoy and Matsu, five miles from Amoy harbour, surely less than Clark Field in the Philippines, or to Japan, is further substantiation. If China is going at some time to have to risk her own destruction for some mimimal target, it is probable that she would try to do some more damage to her opponent than to destroy one of its Pacific bases, which might under circumstances critical enough even double as a 'bait' for just such purposes. There has been advice in the past at various levels for a 'pre-emptive strike' against Chinese nuclear installations. A reply by the United States to a prior Chinese attack would be a situation under which Soviet response even under the Soviet–Chinese mutual defence treaty would not be mandatory.

Why would China 'offer' to attack Japan or India? What would she gain thereby? How would that threaten the United States ('keeping commitments')? What claims has China against Japan, or could she even manufacture any? Japan is the closest and potentially strongest military power China would face in Asia with the strongest industrial base. It would do China much better to attempt to wean Japan away from her military associations with the United States and keep her within her present non-nuclear treaty limitations rather than give Japan its own reasons for reinforcing such military associations or making them permanent. There seem to be no reasons for threats to Japan to be any more likely or plausible than threats to the United States and if anything to be far less so.

Rather than enter into a thorough discussion of the possible role of nuclear weapons in Chinese strategy, which is amply dealt with by Hsieh and Halperin (each of whom has written respectively a book and a long series of papers), and others,[16–20] it seems more useful at this juncture to make some examination of the question of whether China is aggressive or not. Some comments concerning this are now offered; these cannot of course be a thorough assessment.

### *General Remarks*

A few short quotes from Fairbank will first be given to bridge the period of 1946–66; a period which includes such critical stages in United States/China policy as the 1949 White Paper, Korea, Formosa, Quemoy and Matsu:

> Yet the Chinese people wanted peace, the United States had enormous prestige and power to bring to bear upon the scene, the Communists stood to gain from the agreements, the Kuomintang was heavily dependent upon us. Not the least of our problems was how to get a party dictatorship to pursue democratic reforms in order to head off a revolution. Actually we had a divided objective: to press the Kuomintang leaders into reform which would diminish their autocratic power and facilitate internal peace; at the same time to strengthen the Kuomintang-controlled regime as a step toward political stability in East Asia. We became involved in continuing to build the Kuomintang dictatorship up materially at the same time that we tried to get it to tear itself down politically. But we could not control Chiang and his generals, who preferred to do things their way. They relied on their new arms. . . . At all events, we have had a prolonged national distaste for the topic of Communist China. For example, we have supinely acquiesced in a blockade of elementary news reporting, as though we could put our faith in a combination of our mighty Seventh Fleet and our even greater ignorance. If it be true, as I am sug-

gesting, that our emotional reaction to the rise of Chinese Communism has been an attempted rejection of reality, the reasons for this escapism are not far to seek.[20]

It is not possible here to study the China policies of the various United States Secretaries of State since 1945, and, more particularly, of the American positions once the 1945–52 generation of China experts were gone from the State Department. However, one recent example of the kind of advice the administration has been apparently most willing to take concerning China can be given. Presenting testimony before the House Foreign Affairs Committee, 27 January 1966, Charles Wolf, Senior Economist of the Rand Corporation, stated:

> I am dubious that China's fears of encirclement are going to be abated, eased, relaxed in the long-term future. But I would hope that what we do in Southeast Asia would help to develop within the Chinese body politic more of a realism and willingness to live with this fear than to indulge it by support for liberation movements, which admittedly depend on a great deal more than external support . . . the operational question for American foreign policy is not whether that fear can be eliminated or substantially alleviated, but whether China can be faced with a structure of incentives, of penalties and rewards, of inducements that will make it willing to live with this fear.

The point is further clarified by Thomas Schelling at the same hearings:

> There is growing experience, which the Chinese can profit from, that although the United States may be interested in encircling them, may be interested in defending nearby areas from them, it is, nevertheless, prepared to behave peaceably if they are.

Let us in contrast look at the American perception of the Cuban crisis:

> . . . 'The first imperative must be to deal with this immediate threat, under which no sensible negotiation can proceed.' . . . The continued building of missile sites, which would be operational by the following Tuesday, was of even more

> concern. Theodore Sorensen, speaking of the events of October 27, said 'Obviously these developments could not be tolerated very long, and we were preparing for a meeting on Sunday [October 28] which would have been the most serious meeting ever to take place at the White House' . . . Kennedy said that he attached great importance to a rapid settlement of the Cuban crisis, because 'developments were approaching a point where events could have become unmanageable'. According to one source, all agreed that the Soviet missiles had to be removed or destroyed before they were operational; thus, an air strike against the missile sites was planned by no later than Tuesday, October 30.

Through my professional training I have a good knowledge of 'conditioned emotional responses' and 'conditioned avoidance responses', as they are called in classical psychology. Unfortunately for Mr Wolf, and the peace of the world, nations are neither pigeons nor rats. Several parameters of human behaviour, of nations, of leading élites, and of the maintenance of power bases, intervene: not the least of which are various derivations of vanity and 'humiliation'. 'Inequity' and 'injustice', though vague phrases, are powerful irritants to humans, and may cause individuals and nations not to comply with the 'schedule of reinforcement' prepared for them by others. If Soviet missiles in Cuba were perceived by us as a threat to the United States why should not our missiles be perceived as a threat to the Chinese? If Halperin can describe our 'vital interests' as being on China's borders, are not China's 'vital interests' at least also at China's borders, if not by that excellent logic, at our borders? If Halperin can recommend action when our 'vital interests' are not at stake, in anticipation of the day when they will be, when and where should we not act, and what prevents that theorem from applying to all nations about the earth? The whole series of juxtapositions between reality and claimed perceptions is very unpleasant and the degree of optical aberration in our myopic filter required to see things only one way is very great. The witticisms about the 'Chinese Seventh Fleet' off Catalina Island, Puget Sound, Nantucket, or Gardiner Bay are old and stale, yet they are Chinese and American political and diplomatic reality. The

Winter 1965 issue of the English language Taiwanese journal *Vista*, sent to the United States, has two pages of illustration of Taiwanese forces supposedly creeping up mainland China beaches. The author cannot corroborate the claims, and does not know where the photos were taken, but air raids and commando unit sabotage on the mainland are recorded in American newspapers. The point is not that these are minor and petty, which they are, and that the description is irrelevant to one's judgment of which side should rule China and best represents freedom. The point is that we would not tolerate it if done to us, and that it occurs there only and directly because of our sufferance, co-operation and support. If China opens an African consulate, it is broadcast as a new threat and subversion. Perhaps it is. However, we have consulates at all the same places, and have had them there longer, and diplomatic history does not substantiate that the Chinese will do through their consulates anything that we and other nations do not do, whether it be supply of weapons or training local military units, etc. There are two separate strands here. One is containment. What are the results of containment? We have tried 'Cordon Sanitaires' before, and we did not recognize the Soviet Union from 1917 to 1933, which is still being decried by at least some Sovietologists in specific reference to the paranoid effects it may have provoked or abetted in Soviet behaviour. Now we are containing China, and the non-recognition lapse is 1948–52 to 1968. If China is 'paranoid' now, it seems a merited and justifiable paranoia. The second theme is that 'we the [United States] are a Pacific power', that it is our duty to make the 'Asian peace'. In 1936, on the eve of hostilities in North China, the Japanese stated their Basic Principles of National Policy. These included the use of moderate and peaceful means to extend her strength, to promote social and economic development, to eradicate the menace of Communism, to correct the aggressive policies of the great powers, and to secure her position as the stabilizing power in East Asia. Of course, even in 1937, the Japanese government had no territorial designs upon China. The superficial parallels are too close not to be a bit uncomfortable. On what basis we deem ourselves competent to set the terms and fashions of accommodations between Asian nations, when our view of what the proper response to the problems set by China's rise as a world power and her political philosophy is essentially different from that of India, Japan and Indonesia, and also from that of some of the smaller Asian nations, remains a question. The Asian countries that might technically, now or in the future, be able to withstand China in a military conflict are India, Indonesia, Japan and Australia. India, Australia and Indonesia could presently be considered 'anti-Chinese', in opposing Chinese influence and growth. The Japanese seem to show less fear of their neighbour than we think they should be imbued with, on the other hand the foundation of a new nine-nation Asian political union, excluding the Chinese, whose direction and purpose is still not yet set, was formed through Japanese efforts. The nation that claims military and territorial prerogatives (disposition of forces and armaments, patrolling, over-flights, etc.) that it allows no other nation on earth, that nation in history is most suspect in its claims. The power disposition after World War II, and the nature and rise of communism notwithstanding do not alter the historical lesson. The claim that events and requirements of these twenty years are somehow essentially different from those of preceding years is dubious.

*Perspective – Other International Conflicts*

Without going into details on the elements, numbers and positions of the United States Seventh Fleet, the previous remarks have had some reference to the actual disposition of forces of the United States relative to China. Let us look at some aspects of the verbal fray. In order to better fit into perspective official American estimates of China's behaviour, which will enter into the following discussion, let us look at a few armed conflicts now going on or only recently concluded around the globe.

(*a*) Kurd–Iraqi; intra-national, estimates of casualties for five years of combat reach 300,000; Iraq is a member of the United Nations.

(*b*) Pakistan and India; international, 1965. Both nations are members of the United Nations.

(*c*) Kikiyu–Masai (Kenya) nad Watussi–Bantu (Uganda, Ruandi); intra-national mortality estimates 50,000 in each case; both nations are members of the United Nations.

(*d*) Indian–Naga; intra-national, eight years' duration, estimates of casualties, 50,000. India is a member of the United Nations.

(*e*) Egypt and Yemen; international, two years' duration, casualties 50,000+. Egypt is a member of the United Nations.

(*f*) Serious intra-national conflicts with large casualty rates and of several years' duration exist in several Latin American states. *One* of these alone, Bolivia, has run up a ten-year mortality figure of 200,000. All are members of the United Nations.

(*g*) Indonesia, recent civil war, estimates of mortality figures as high as 300,000–500,000 are probably correct. Though Indonesia had withdrawn from active participation in major United Nations affairs, had she not, it is improbable that American officials would have called for her expulsion.

(*h*) Arab–Israel conflict, international, over ten-years' duration. All participants are members of the United Nations.

(*i*) Nigeria; intra-national, a year's duration, 200,000 deaths. Nigeria is a member of the United Nations.

(*j*) United States post-World War II armed actions in the Dominican Republic, Cuba and Vietnam are not quite looked at throughout the world in the way the United States administration looks at them. The United States is a member of the United Nations. The same situation of course holds for Soviet intervention in Hungary and Czechoslovakia. The Soviet Union is a member of the United Nations.

The point of these examples is simple and obvious. There have been several international conflicts since World War II (the examples listed are only a small fraction), with large numbers of casualties and dead, with real consequences, and some in which 'aggression' was even relatively obvious and easy to attribute to one of the parties. However, in no case has the United States interpreted any of these as a threat. No major competitor to United States power, policy or aims was directly involved in any case. We have not sought to invoke against any of these conflict participants what we seek to invoke against China on poorer evidence, nor claim to any of the participants any relevance of the conflicts to their competency as United Nations members, either as evidence for or against.

Let us turn now to the armed conflicts that China has been involved with:

(*a*) *Taiwan–Formosa.* Formosa had been ceded to Japan in the Treaty of Shimonoseki in 1895 as part of the peace settlement reached after the Chinese defeat in the Sino-Japanese War of 1894–5. Prior to that time Formosa had been loosely associated with the Chinese empire, having been formally annexed in 1683 by the Ch'ing Dynasty. During this long period of Chinese affiliation, the island enjoyed considerable autonomy. In the Cairo Declaration of 1943 and the Potsdam Declaration of 1945 the Allied Powers in World War II proclaimed their intention to transfer Formosa from Japan to China. The Nationalists thus recognize Formosa as part of China, as indeed they claimed before 1943 and the Cairo Declaration and as they also recognize Tibet. They claim the mainland as part of China that they do not hold and which they hope to reclaim. At the very least this makes logically difficult the American position that the CPR cannot claim Formosa as part of China which *it* must still reclaim, and would if the United States did not prevent her from doing so. It makes it equally difficult to logically claim that we are supporting the Nationalist position. That all sides recognize this, or did in the past, is implicit in the attitude of the United States to the CPR and the problem of Formosa just before the outbreak of the Korean war.[22] Some have felt that United States recognition of the CPR was then near. In the main it was expected that Formosa would join the CPR relatively soon. Discussions between the U.S. and the CPR concerning Formosa have taken place subsequently[23] but in a less realistic context. At any rate it is obvious that the bargaining position of the United States is military and political, and not legal, and that labels of 'aggression' with regard to China's intentions concerning Formosa are inappropriate.

(*b*) *Korea*. In 1949 the Chinese revolution ended. In 1950 the Korean war started. At the time North Korea was considered under Soviet influence and not under Chinese. The Chinese were not ready, eager or planning to do anything in Korea, which is the same as saying that it is difficult to attribute the action to their provocation or proxy. Aggressions are seldom impromptu, unplanned and unprepared. (This is *not* to imply that the outbreak of hostilities in Korea should be presumed to have had Soviet support either. The evidence is strong as well that though the crisis was building up in North and South Korea for some weeks, both major powers were taken by surprise by the actual events, the Societ Union the more so of the two. The outcome of the events in the United Nations in the several days immediately following the outbreak of hostilities had much to do with setting the framework for subsequent interpretations of the war.)

> The weight of evidence is against any serious Chinese involvement in the start of the Korean war. Relations between Peking and Pyongyang do not seem to have been close. It was not until August 1950 that an ambassador from Peking presented his credentials at Pyongyang, and there is evidence of earlier disputes between the North Koreans and the Chinese Communist authorities in Manchuria, which were resolved only by Soviet mediation. The emphasis of Chinese Communist publicity was on the conquest of Taiwan and Tibet, and Chinese troop dispositions appeared to be primarily designed for these objectives. . . . The material from the interrogation of prisoners shows that even the forces in Manchuria received little preparation for intervention until shortly before they crossed the Korean border in October.[23]

Whiting's book goes further to indicate the reasons why China eventually and reluctantly entered the Korean conflict.[24] China felt that it had made clear that it would not intervene if the 38 parallel was not crossed. Whiting considers the Western discrediting of this intention due to the means by which China had transmitted it, in Lord Lindsay's words 'not to take them seriously', a failure in communication. The American government has never failed to refer to 'Chinese aggression in Korea'. The explanation is inadequate and inappropriate.

(*c*) *Quemoy and Matsu.* Quemoy and Matsu are five miles from Amoy Harbor, and at no time before 1949 did anyone ever doubt that they would be under mainland China's control. They traditionally fall under the jurisdiction of Fukien province.

(*d*) *Chinese and Indian border disputes; The Aksai–Chin and Ladakh and The McMahon line.* To begin with, the McMahon line has been in dispute since the day it was drawn by the British Colonial office. China refused to ratify it within two days after the Simla Convention in 1914 initialled it. Britain stated that it would sign independently in default of China's adherence, and would regard the convention concluded thereby. World War I broke out, and the matter was forgotten by the West. Since it was never drawn by Indians and Chinese in conjunction, it is not beyond imagination that there might arise difficulties over it once both these nations achieved independence. (Similar lines drawn by the British between Kenya and Uganda, Kenya and Somalia, and Uganda and Sudan are now also in contention, armed contention, and the reported casualties resulting from strife on those borders seems about ten times as great as that reported for the China–India conflict. For some reason, there have not been great tirades about this in the American press, nor have these countries been asked to give up their seats at the United Nations.) Border disputes between China on the one hand, and Pakistan, Burma and Nepal have all been settled by negotiation, though these too have received little or no press notice. Lall states that 'Nehru was not in a negotiating mood with China over the border dispute for several years before the Chinese took military action in the autumn of 1962. To the Chinese, Nehru looked downright truculent', and in a more recent publication, that it could not be assumed that 'Peking alone is to be blamed for the armed clash that occurred. The situation was far too complex for so simplistic a summing up.'[25] As the Indian representative to the United Nations, Lall must be considered an informed participant, if not at least an informed observer, of these events. Nevertheless it has been claimed that secret negotiations were taking place between India and China, that proposals for exchange of territory in the Aksai–Chin area (to go to China) for territory in the McMahon line area (to go to India) were being discussed, and that due to the strength of feeling on these matters in the Indian parliament, press and public, it was necessary to keep the negotiations secret until they were completed. Then the Tibetan crisis broke, the Dalai Lama fled to India, and ostensibly the negotiations broke down. Since this version depends on unpublished information, there is no way to assess it, or to credit to either side. The questions as to why India did not indicate any claims in the Aksai–Chin during the five–six years in which the Chinese had apparently notified her that the road building was being carried out may be answered by the above. Why the incidents and military engagements occurred finally when they did, and whose troops it was that were immediately responsible for the firing, remains to be answered. Though the word 'political' is not an antonym of 'aggressive', and though all things are in some sense 'political', and though in other instances, a situation might in fact be correctly categorized as both political and aggressive, it might be more informative to describe China's border dispute with India as 'political' rather than 'aggressive'. Tibet enters strongly into the situation. Why was Pakistan able to make a border settlement with China which Pakistan considers to be generous? Why was India not able to do so? The fact that it was not, strongly implies that it is more than the border problem that is at issue between the two. It is rare in the history of territorial and border disputes that a nation in armed possession of some amount of territory will withdraw from it though not under military pressure to do so, as the Chinese did in the latter stages of this conflict. Critics of China's position must find a hard and realistic explanation for this withdrawal consistent with the other terms they apply to Chinese international goals. Though it may not have been

China's intention to do so, an argument could even be made, based on this action, to claim that the incident could be taken as proving to other Asian nations the irrelevance of Chinese military threats. Astute commentators have also offered explanations for China's action in terms of its intended effects on the Sino-Soviet dispute, as well as on India's domestic politics. It is impossible for this author to assay these interpretations.

(*e*) *Tibet*. Apparently there is a 1939 United States State Department declaration recognizing Tibet as a part of China, and this was restated in the 1943 Cairo Declaration. Whatever else it should be called then, 'aggression' is not the right label for what occurred. China had held 'suzerainty' over Tibet on and off for a thousand years. The British Raj in India had made a policy of hoping the problem of Tibet and China would go away, that is, that China would never become a power capable of realistically indicating her interests in the area. Much of the British tradition of concepts dating from 1890 to 1915 concerning the security of the Indian subcontinent were taken over by India as the basis for her national security after 1947, as one clearly sees from her relations with Bhutan and Nepal. It is beyond my capabilities to explain how the historical process in *Tibet-Chinese* relations came to be viewed by *India* as a 'serious threat to Indian security'. A cursory look is confusing to a Westerner, as the area clearly did not have the European example of mutual invasions of the last 300 years.

> The main reason for British involvement in Tibet arose from the need to establish a security system in the north-eastern Himalayas in order to protect the Indian subcontinent from any possible threat of Russian power and influence in Tibet. Imperial China did not at the time represent an immediate threat to the British Raj in India. Therefore, beginning from 1906, Britain persisted on China's suzerain rights over Tibet, and even got Russia to recognize and abide by it.[23]

Suzerainty is a difficult concept for the modern world, and one would not have expected it to last long. It would go one way or the other, to complete independence or complete incorporation. Tibet participated in the Chinese Nationalist Assemblies in 1946 and 1948. The history of Tibetan–Chinese relations before 1900 and from 1918 to 1948 would also have to be assessed to make a judgment, but the case comes closer to those marked 'intranational' rather than 'international'. There has been some attempt at objective discussion as to why the revolt and fighting occurred when it did. Land reform; i.e. legislation regarding it, was instituted in China in 1949. Other sources indicate that the terms of the legislation did not become incumbent upon Tibet until May 1951 and the signing of the Sino-Tibetan Treaty. A ten-year postponement for meeting its terms was given Tibet. The fighting, and entry of Chinese troops, when it occurred in 1959, apparently was over this issue. This is never mentioned or discussed, even to be disputed, by White House or State Department spokesmen. One is reminded of the Indian 'reclamation' of Goa, or the Nationalist Chinese occupation of Formosa. All were 'settled' by armed interference. In addition Fairbank describes operations carried out in 1947 by the Nationalists arriving on Formosa in which several thousand leading native Taiwanese were killed.[21] Communist Chinese relations with Tibet have figured prominently in public estimates of China's behaviour by the very highest administration spokesmen in recent years, and are described in most horrendous terms. There are few essential differences between that episode and our treatment of the American West and its then inhabitants. In fact, in our historical instance there was not a thousand years of previous control or suzerainty as precedent. In addition numerous European powers carried out far more extensive programmes, innumerably more costly in native life and manners, in territory not contiguous with their borders, for centuries of their recorded history. Western criticism seems slightly self-indulgent, if not hypocritical. There are numerous and obvious, specific factors of historical time and place that make Tibet and the American West non-analogous, and perhaps make criticism more valid in the light of today's opposition to repressive tendencies, but it is equally obvious that the models the United States has sought to fit to Tibet are inappropriate.

*General Remarks*

For some ten years now, at each instance of sponsorship of Chinese membership in the United Nations, Ambassadors Lodge, Stevenson and Goldberg have reviled China's 'aggressiveness', 'warlike disposition', 'aggressive intent', 'deep expansionism', etc. The border episode with India gave American spokesmen, in and out of the United Nations, new phrases and impetus for their arguments. This kind of name-calling on an international level, together with the influence the United States has been able to marshal to effectively ostracize China diplomatically, is not a petty or benign thing, especially if it happens to concern the most populous nation in the world, which simultaneously feels that it is for the first time capable of entering and entitled to enter the modern world scene. It may be assumed to so specifically provoke from China in response the kind of behaviour that we claim to decry – their penetrations in Africa and Latin America – that our expressions of horror at these latter efforts on China's part can be doubted. It is only to be expected that China will apply pressure where it can if the United States applies pressure where it can. Undoubtedly the Chinese verbal fomentations are far cruder than our own and decidedly unpleasant to our ears. However, China has lacked two things to place alongside the rhetoric, military power and influence in the United Nations. Without these their name-calling takes on a rather different functional aspect than our much more temperate language. In addition the past, which we never make proper use of, has much to say about methods that would directly alleviate their improper language. What Mr William Bundy calls 'errors in justice and conduct in our relationships with China', are not correctly relegated to pre-1900 history.

And now the Chinese have presented us with another stick with which to beat them. Mr Cyrus Vance said:

> Exactly five weeks ago tonight – in a distant country – a political document appeared. It was written by Lin Piao, the Vice-Chairman of the Central Committee of the Chinese Communist Party, Vice Premier of Red China, and the Minister of National Defense.

> I want to talk to you about that document tonight. For it tells us not only what Peking's intentions are in Asia – not only what Peking's intentions are toward the United States – but what Peking's plans are for the whole expansion of world Communism.[26]

There are several other readings that can be given to Lin Piao's statement of 3 September 1965, and, very noticeably, it is such other readings that have been put forward by nearly all China experts in the United States. Like so many communist documents since 1917, it is a combination of analysis and prognostication, and probably resembles a good many United States State Department analyses in these formal aspects. It seems anything but a 'plan'. The rapid categorization of the document by the President and the Secretaries of State and Defense as a 'Chinese Mein Kampf' seems a deliberate and energetic public misreading. Mr McNamara some years back even attempted to make use of such interpretations in bolstering the flagging determination of the NATO ministers, as he presented them with a tripartite vision of China: as aggressive and expansionist; Lin Piao's document as 'proof'; and finally China soon to be nuclear armed and with missile capability. The major interpretation of Lin Piao has in fact been that the message to communist movements in Asia was that revolution was *not* directly exportable:

> In order to make a revolution and to fight a people's war and be victorious, it is imperative to adhere to the policy of self-reliance, rely on the strength of the masses in one's own country and prepare to carry on the fight independently even when all material aid from outside is cut off. If one does not operate by one's own efforts, does not independently ponder and solve the problems of the revolution in one's own country and does not rely on the strength of the masses, but leans wholly on foreign aid – even though this be aid from socialist countries which persist in revolution (i.e. China) – no victory can be won, or be consolidated even if it is won (Lin Piao).

It was in the period 1960–2, in that instance in specific reference to Vietnam, that China developed its argument on 'national liberation' wars, which was to become so widely publicized in 1965. In the Vietnamese case the argument could be put as:

> . . . the Vietnamese themselves must develop local political support. If outsiders undertake to aid the struggle of the southern Communists, the latter will not need to rely on their own people and resources. In revolution, violence alone can cement the bonds between the guerrillas and the peasants. When the Vietnamese themselves have met the test of violence, they will have forged an unbeatable popular backing.[27]

Moreover these ideas were not novel in Chinese political theory:

> Even as early as 1950, therefore, the foreign policy uses of the 'path of Mao-Tse-tung' did not imply a general call to the Asian Communist parties urging them to attempt the Chinese tactics of armed struggle. Indiscriminate partisan warfare had already been encouraged by the Cominform and it had, as the Chinese well knew, resulted in the total defeat of nearly every Asian Communist movement. By 1950, the Maoist path, in fact, amounted to a basic criticism of ill-prepared armed struggle tactics and an alternative program calling for the complete reorganization of the battered Asian parties.
>
> It was precisely between 1955 and 1958, when Peking took a moderate foreign policy stance ('The Bandung Peaceful Coexistence Policy') that the United States initiated most of those actions in Asia which subsequently did so much to create growing hostility between the two powers. In point of fact, Washington was then (and still is) intensely worried about the possibly adverse consequences of accepting a genuine 'peaceful co-existence' posture in Asia and therefore set about to insure that there would be no adjustment to, or compromise with, China which might heighten her prestige and influence. The United States did not like the idea of 'peaceful transition' either, since it had good reason to doubt how long some of the elites it supported in various countries would survive real 'free' openly contested 'democratic' elections.
>
> It is against this background that China's changing emphasis on, and interest in, supporting revolutionary move-

ments must be evaluated. The strident Chinese call to 'revolution' did not come in 1957, or in response to Soviet missile and nuclear technology. It came in 1960 after a long series of provocative American actions in Asia which Peking saw as invalidating the long term viability of a 'soft line' Bandung policy. Persistent American pressure, coupled with Soviet disinterest in curbing that pressure, played a far more determining role in Peking's decision to support revolutionary movements than any sudden euphoric confidence in the prospects for 'people's wars' and 'national liberation struggles'. These revolutionary forces, whether led by Communists like Ho Chi Minh or nationalists like Sukarno, began to acquire major strategic importance to Peking once it became clear that a policy based on alignment with Russia and cooperation with the Afro-Asian neutrals could not be manipulated, effectively, to weaken or deter the United States. Since Peking was not prepared to accept a determination of the status quo in Asia unilaterally imposed by the United States, her previous strategy had to be altered. Peking's actions after 1960 indicate that this alteration took two forms: first, through a sustained ideological assault on Moscow's stance in the Cold War, China attempted to force the Soviets to abandon their 'erroneous' line. Second, while maintaining friendly relations with genuinely neutral states, Peking began to lend vocal and, after 1963, material support to any 'revolutionary' movements deemed capable of weakening or embarrassing the United States, whether or not such movements were led by Communist parties. This approach cost China very little since the revolutionary activity, though indirectly supportive of Peking's strategic goals, was done by others and consequently entailed no serious risk of a direct U.S.–China confrontation.[27]

What was the result of China's policy?

What Chinese policy had accomplished was to incite an even more militant and active policy on the part of the United States in Asia. Not by deeds, but certainly by words, China had supplied ample evidence of revolutionary zeal to support the arguments of those quarters in the United States which felt that a decisive battle must be fought in Asia to 'contain' China, to discredit her thesis on revolution and on how to deal with American power. Where Peking had totally failed

> to convince Moscow that the decisive arena was the 'national liberation movements' struggle against 'imperialism', she succeeded in convincing the United States.[27]

With the development of Chinese nuclear weapons, the United States immediately began to speak of the Chinese nuclear threat, and of nuclear blackmail. Nuclear weapons have long been based on Taiwan, and on American ships off the Chinese coast. By 1960, Matador missiles had been installed on Taiwan and these were capable of delivering nuclear warheads against the Chinese mainland. President Eisenhower has indicated that there was readiness to use nuclear weapons by the United States at the time of the Quemoy and Matsu crisis, and there are more detailed indications that the required ordnance was landed on the island during the crisis.

At the time of the Pueblo crisis one learned with irony that carrier based planes in the Sea of Japan were only fitted for delivery of nuclear weapons. Nevertheless, our traumatic apprehensions are used to make contemporary national policy, which will have repercussions for many years to come, all the longer because strategic nuclear weaponry is concerned. There is value here in something that has come to pervade the understanding of researchers in many fields of science. The constructs one uses set conceptual limitations, largely through the role of language and by excluding ideas one will not entertain. This realization is evident in the writings of some classical strategists.

> One could go on with endless examples to demonstrate what should hardly need demonstration, that bias is not easily curbed, partly because its possessor usually does not recognize it as such; that bias affects and conditions perception as well as action, and the effects of bias on perception are not desultory or spasmodic but constant.[28]

If further substantiation were felt necessary there is much to be said here for the application of knowledge concerning expectations and their effects on resulting behaviour and interactions derived from research in psychology. It is possible to think of few situations in the world to which this is now more relevant than the American attitude to China. What is worse is that the constructs we use for China seem to be more than accidentally deceptive or misleading. They are purposefully false, and are perpetuated in that form.

The emergence of China as a nuclear power is important for its political and psychological implications. It is relevant that we have had ten years in which to prepare for this situation, which have not only been frittered away – which would be a mildly passive response – but years in which we have actively set up an environment concerning our relation and understanding of China, which make any but a limited and maladaptive group of responses now very difficult, if at all likely. Particularly non-functional and maladaptive as a political and psychological response are ABM weapons. As for the future 'larger military threat', wars require two countries – that is, their genesis is *most* often bilateral. An American study of the Chinese view of the military threat to her, as American nuclear power developed directly off her shores from the 1950s to the present might be relevant and functional in avoiding a projected conflict.

We have been looking at the assumptions that surround or underlie discussions of whether the United States Sentinel ABM system can be considered anti-Chinese or not. For this discussion one's general assessement of the technological capabilities of a particular ABM system are largely irrelevant. Given the announced context of the American decision, what may be its subsequent effects? It is likely that Chinese missile development decisions will only be affected in their details. It is not likely that China will be deterred from completing to the best of her ability whatever stages in weapons development she had already intended. The United States had repeatedly stated that it would have to go ahead with an ABM system unless the Soviet Union would agree to discussions concerning such systems. It initiated an ABM system, saying that it was aimed against China, and then obtained Soviet agreement to the talks about the same ABM system. If domestic political pressures motivated the American decision and Sentinel was neither an 'anti-Soviet' system, nor an 'anti-Chinese' system, some have argued that it would in fact be desirable if it were considered anti-Chinese rather than anti-Soviet, as the Soviet Union has more immediate capabilities to respond in that way which would accelerate the arms race. However, once the Sentinel system is in existence, it will be an anti-Soviet, or an anti-Chinese one, whatever reasons it may have been initiated for. Others have argued that the rationalization chosen in the United States was particularly unfortunate as it will maintain and accentuate the stereotyped images of American–Chinese opposition and recrimination, distorting perceptions further. Though there may be short-term strategic arguments in its favour, such as those expressed by Mr Nitze, the larger military effect would be to push on the confrontation. Some have pointed to the utility of an American ABM system in reinforcing the credibility of the nuclear umbrella for Asian nations. This argument has been stressed in relation to India, and in relation to discussions of proliferation. Others have replied that crediting the Chinese with provoking an ABM necessity will only accentuate the Chinese bomb in particular, and nuclear weapons in general, in Asian judgments. Some have said that the implication that the United States needs protection against the Chinese runs the risk of 'going to their heads', leading China to overestimate the effect of her possession of a nuclear weapon. Hsieh has pointed out that the Chinese seem quite capable in general of making their own reality estimations, and are not likely to be so easily misled.

It would not be surprising, however, if the attendant publicity required in the United States to accompany the decision will act to convince large segments of the American public that China is so dangerous that they need protection from her, and that a threat is realistic. Again, if the American decision was mediated by domestic politics, it might imply that there was no intention to go farther to a larger ABM system. Arguing against this are the implications of Mr McNamara's repeated warnings, the political inability to resist pressures so directly connected with problems of national security, and the vain effort of the move as a domestic political gesture. The decision, particularly in conjunction with the MIRV potential scheduled in the United States for 1973, may have implications for ages to come. At the least, within 5–10 years it will become obvious that we have been saddled with an expensive, strategically difficult and politically hard to reverse ABM system.

There are other aspects of the interaction between the topic of a comprehensive test ban and Chinese foreign policy, such as its effect on Sino-Soviet relations.[29] China is also the only nuclear power to have made a 'no first use' pledge. Obviously this is consistent with China's nuclear capability *vis-à-vis* the West. At the same time it also serves as a direct message to its Asian neighbours as well as perhaps offering the West a lever which it must be imaginative enough to make use of in some way for purposes of negotiations.

We have gone from a discussion of the United States ABM decision as regards China to what is essentially a plea for 'single standardism' on the international political scene. This plea is of course not terribly novel.

> China's exclusion from the U.N. and, in reality, from all international negotiations, particularly in Asia, is creating an unreal situation, which doubtless explains the disturbing reactions on its part that we witness periodically and that, in actual fact, prevent any progress towards a settlement. In Asia as elsewhere, there are specific questions for which solutions should be found. At the same time, there is reason to seek an overall balance of forces and positions, which actually is the problem of peace. The point is to ascertain how this balance can be established all around the huge Chinese empire. . . . A unilateral policy of containment does not seem to be the right answer. What is needed is solutions reached through agreement, or at least resulting from some modus vivendi. . . . The sooner it is possible to begin the better it will be for eveyone.[30]

Since nothing has been done to begin, it is necessary to repeat the plea.

## *References*

1 *The Chinese Question and ABM Deployment*, J. I. Coffey, Study paper no. 6, Office of National Security Studies, Bendix, Sept. 1965.

2 *U.S. Armament and Disarmament Problems*, Hearings, Subcommittee

on Disarmament, Committee on Foreign Relations, Feb. and Mar. 1967.
3 *Aviation Week and Space Technology*, 12 Aug. 1968, pp. 77.
4 'The Case Against An Antiballistic Missile System', J. B. Wiesner, *Look* magazine, 28 Nov. 1967.
5 Editorial, *The Times* (London), 19 Sept. 1967.
6 *The Weapons Culture*, R. E. Lapp, 1968. W. W. Norton.
7 'Keeping the Strategic Balance', *Foreign Affairs*, C. Kaysen, **46**, no. 4, July 1968.
8 *Scope, Magnitude and Implications of the U.S. Antiballistic Missile Program*, Hearings, Subcommittee on Military Applications, Joint Committee on Atomic Energy, Nov. 1967.
9 'The U.S. and Communist China', *Seato Record*, **5**, no. 2, W. Bundy, Apr. 1966.
10 'Nike-X Deployment May wait for Chinese Launch Pads', *Space/Aeronautics*, **44**, no. 5, Oct. 1965, p. 14.
11 *Strategy and Arms Control*, T. C. Schelling and M. H. Halperin, 1961, 20 Cent. Fund.
12 Address by Sec. of Defense McNamara. Before the New York Economic Club, 18 Nov. 1963, Dept. of State Bulletin, 16 Dec. 1963, pp. 914–21, and *Documents on Disarmament*, 1963, pp. 583–94.
13 *Review of a Systems Analysis Evaluation of Nato vs. Warsaw Pact Conventional Forces*, Report Committee on Armed Services, U.S. House of Representatives. Section by Alain Eithoven, Sept. 1968.
14 Authorization for Military Procurement, R & D, Fiscal Year 1969, and Reserve Strength Hearings, Committee on Armed Services, U.S. Senate, Feb., Mar. 1968.
15 *Chinese-Soviet Treaty on Co-operation and Mutual Assistance*, 14 Feb. 1950. There was no time limit set to the treaty, and it is still valid, and has not been renounced by either signatory. The last official Soviet statement in this regard was Mr Khrushchev's in 1963: 'Soviet Union is always ready to come to the military assistance of China. Should she be attacked . . . and . . . it is not important for the defense of socialist countries for two socialist countries to have atomic weapons.'
16 *Communist China's Strategy in the Nuclear Age*, Alice Langley Hsieh, 1962 (Prentice Hall).
17 *China and the Bomb*, M. H. Halperin, 1965 (Praeger).
18 *Impact of Chinese Communist Nuclear weapons Progress on U.S. National Security*, Report of Joint Committee of Atomic Energy (JCAE), July 1967.

19 *Communist China and Arms Control*, M. Halperin, and D. Perkins, 1965.
20 *Communist China and Arms Control, A Contingency Study, 1967–1976.* Hoover Ins. on War, Revolution and Peace (for ACDA), Feb. 1968.
21 *The U.S. and China*, J. K. Fairbank, 1959 (Viking).
22 *The China White Paper Vol. I and II. Aug. 1949*, 1967, Stanford Univ. Press; *United States Relations with China, with Special Reference to the Period 1944–1949.* Dept. State Publication 3573, Far Eastern Series 30, 1949.
23 Review of *China Crosses the Yalu.* Lord Lindsay of Birker, *Science*, 10 Feb. 1961.
24 *China Crosses the Yalu, the Decision to Enter the Korean War*, A. S. Whiting, 1960, Macmillan.
25 *How Communist China Negotiates*, A. Lall, 1968.
26 *SEATO Record*, C. Vance, IV, no. 6, Dec. 1965.
27 'The Maoist Imprint on China's Foreign Policy,' D. Mozingo, in *China Briefing*, F. Armbruster, J. W. Lewis, D. Mozingo, and Tang Tsou, 1968 (Univ. of Chicago).
28 *Defense Policy and Total War*, B. Brodie, Daedalus, **91** American Foreign Policy, Freedoms and Restraints, Fall, 1962.
29 'The Nuclear Test Ban and Sino-Soviet Relations', W. C. Clemens, Jr., in *Sino-Soviet Relations and Arms Control* (ed.) M. H. Halperin, 1967, MIT Press.
30 Maurice Couve de Murville, French Minister of Foreign Affairs, 20 Oct. 1965.

# SECTION D: *General Implications*

M. LEITENBERG

# 1: Civil Defence Programmes; Research in the United States Concerning the Post-Attack Environment

In earlier discussions of active ballistic missile defence (1960–5), passive defence measures such as fall-out shelters and shelters for protection against the effects of blast and heat were also discussed. There has been a hiatus in the discussion concerning merits, demerits or procurement of these facilities, but not in ongoing research that would be necessary to indicate the requirements of such systems and to support their procurement should the question be reopened. In the 1965–8 period, there has been little or no mention of Civil Defence (CD) in the discussions in the United States of the 'thin' or 'thick' ABM system, or in those which surrounded the procurement of the Sentinel ABM system. The following short item appears in the 1969 United States Department of Defense appropriation hearings:

> *Question* 62: The testimony states with respect to both Postures A and B defenses against a Soviet nuclear attack, 'we would also want to expand and accelerate the fallout shelter programme'.
>
> Why is this sentence left out of the unclassified testimony?
>
> *Answer:* In moving from the classified to the unclassified version of the statement, the phrase referred to was dropped for editorial reasons. The point that an expansion of the civil defence effort would be an integral part of any large Damage Limiting program has been made frequently in prior years and could just as well have been included in the unclassified verson of this year's statement.[1]

In the Senate Committee on Foreign Relations, Sub-Committee on Disarmament, Dr J. S. Foster was asked:

Tell us something of the magnitude and cost of a shelter program to defend the major cities in the United States.

*Answer:* The civil defence program we are now pursuing is estimated to cost around $800 million in the period FY 68–FY 72. This will provide, with what has gone before, about 230 million shelter spaces. An expanded fallout shelter program to accompany an austere NIKE X deployment would cost about $800 million additional and add a total of about 50 million fallout shelter spaces by the end of FY 72.

(Note: depending on the assumed time of an attack, the population will be distributed differently, so that there is not a one-to-one correspondence between shelters and population.)

When civil defence advocates presented for national approval their programmes for both fall-out and urban blast shelters in 1960–1, these proposals ran headlong into overwhelming difficulties and opposition, in both the technical and political arenas. There are very strong feelings among civil defence advocates today not to repeat the same technical and political 'mistakes' the 'next time'. What this determination means functionally is hard to say. There will always be a certain category of items raising uncertainties concerning post-attack human behaviour. These will never be entirely removed no matter what the level of technical development and certainty engineered into the physical aspects of the shelter system. Barring an infinite expenditure and an entire move underground by the strategic powers – of cities, food production and storage facilities, power production, etc., newer strategic delivery systems will continue to attempt to undo any savings derived from the installation of passive (or active) defensive systems. Questions relating to the degree to which a large CD programme would exacerbate the strategic confrontation it ostensibly is attempting to protect the nation from, would always remain. Finally, it is difficult to foresee the criteria that CD advocates anticipate that would alter the present widely assumed United States congressional opposition to a large-scale blast shelter programme.

Because of the mention made in Chapter 1 to the United States Office of Civil Defense budget for 1969 it is relevant to

Office of Civil Defense (OCD) expenditures for previous years and, therefore, these are given in an appendix to this chapter. Also included are remarks relevant to CD which appeared in the most recent statement of United States Department of Defense witnesses before Congress, and rather than intrude these here, have again placed them in an appendix. One can simply summarize these remarks by saying that a rather extensive United States national fall-out shelter programme continues, hoping by the end of 1968 to have identified 170 million spaces, to have marked 101 million, and to have stocked 55 million with an average of 14 days' supplies.

All of the subsequent discussion of Civil Defence will be in terms of research on the post-attack environment. In January to May 1967 the author had an opportunity to read or look at some unclassified reports on aspects of civil defence. Their source and number follow.

| | |
|---|---|
| 1. Rand Corporation | 33 reports |
| 2. Hudson Institute | 42 reports |
| 3. Oak Ridge National Laboratory; Civil Defense Research Project | 18 reports |
| 4. Office of Civil Defense | 5 reports |
| 5. Stanford Research Institute | 22 reports |
| 6. National Academy of Science – National Research Council | 20 reports |
| 7. Miscellaneous | about a dozen reports |

This is an unknown, and probably small, fraction of the total of such reports. Civil Defence related research, for specific shelter 'hardware' items as well as for systems analysis, is funded by the Advanced Research Projects Agency (Department of Defense) (ARPA), the Defence Atomic Support Agency (DASA), the Office of Emergency Planning (OEP), the Atomic Energy Commission (AEC), the U.S. Air Force, the U.S. Army and probably by other agencies, as well as by the Office of Civil Defense. Many other smaller private research corporations in addition to those listed above have such contracts: IIT Research Institute, Research Triangle Institute, Betelle Memorial Institute for Research, Human Sciences

Research Inc., Planning Research Corp. Research is also undertaken by many individual researchers in universities. Some also is carried out in various government departments (Agriculture, Health, Education and Welfare), and by the military services themselves (U.S. Navy and Air Force). At the very least one has to say that there is a great amount of CD related research going on in the United States, though the sums of money involved are very small in relation to most other United States defence related expenditures. Nevertheless, the listing of other agencies that fund such research indicates that it is insufficient to look at the OCD funding alone in attempting to form a correct estimate of the expenditure in this area.

Whether *any* amount of research is capable of 'solving' the problems of a potential post-thermonuclear exchange environment is another question.

How good are the above reports? Some are so inept, and so obviously transfer standard peacetime methods of thinking to the most extraordinary non-peacetime situation imaginable, as to be meaningless. Some make attempts to be thorough and rigorous. However, most of the latter concern material supplies. For example, one set of documents which was found somewhat impressive, probably because they tell more about the complexity of the pre-attack economy than they do about the potential of post-attack recovery, is a group of over fifty technical reports and 187 technical manuals prepared by the National Planning Association for the National Resource Evaluation Centre (NREC) of the Office of Emergency Planning (OEP). These all concern computer programmes containing stored data on various damage assessment programmes, manpower and economic resource surveys, and PARM (Programme Analysis for Resource Management), concerning administrative criteria and goals in the post-attack situation. Here for example one will find pages of material on 'Footwear, excluding Rubber', with estimates of how much would be contaminated and useable, moderately damaged and usable, contaminated and lightly damaged and usable, etc., and estimates of time and resources needed to reinitiate production. In fact, if one needs information on the national resource base, to know what is and what can be made in peacetime, one would go about it as is done. However, the PARM Project Final Report states in its introduction:

> These production schedules are limited by the surviving stock of products and producing plants. About 300 industries are considered separately. If capacity is short in any industry, added capacity is created as required, by decontamination, damage repair, or new construction. Such additions to capacity are, of course, scheduled only after allowing time to perform the needed construction work. Production is also limited by surviving effective manpower of each industry, plus additional manpower which could be trained. In some cases, substitution of one material or product for another is scheduled in order to break a bottleneck.

And now one has the obvious questions: How much decontamination equipment will remain? Will its distribution be random in relation to need? How much will losses in personnel trained to carry out the decontamination procedure affect utility of surviving equipment? With what materials will you repair damage and build new construction? Will such material be deliverable, or will delivery routes and means and energy sources for delivery vehicles be interdicted. Most critically, what one wants to have is an understanding of the 'deficit interactions' that a series of degraded and interacting industrial elements would have on each other, i.e. to what degree does the degradation in a chain have a spreading or 'snowballing' effect: power, parts for power generation, fuel for power generation, parts for transport, fuel for transport, transport for fuel, road and railbed for transport, parts for production machinery, degradation of all the 'parts' production industries; metal and alloy supplies for machinery and parts production, etc. (In some areas, such as communications, hardening of facilities has quietly proceeded to some degree as part of present CD or other federal programmes.) Those who formulated the PARM models are aware of these obvious questions (though at times such awareness seems to be absent in the phrasing of many other studies, despite the obviousness of the questions), and there has been an attempt to answer them by an input-output matrix of 190 units, based on the 1947 data of the Bureau of Labor Statistics. The criticisms that one has of this attempt can be basically separated into two groups. One has the conceptualization for a very sophisticated evaluative technique, with a very poor data base:

(*a*) The input data is too old.

(*b*) The matrixes and the data they contain are too broad, as they are by industry, or even by groups of industries. One simply needs enormously greater detail: i.e. within a particular industry are there certain critical elements; a particular solvent, lubricant, alloy or bearing, and how wide then is its production and distribution base?

(*c*) The post-attack postulated deficits are in terms of a percent loss to the industry. This may again be not specific or particular enough to work from in designing reconstruction programmes.

(*d*) Geographical–regional data seems weak and also not detailed enough.

(*e*) Authoritative economists have informed the author that as detailed before-the-event central planning as is here required is impossible even in peacetime, at any rate at present, because not nearly enough is known about the techniques of production, the way they change, and their interdependence, it is therefore not now used even for peacetime economic planning, and there is no *a priori* reason to suspect that it would function any more satisfactorily in a wartime or post-attack world. In addition there are even questions from within the economic profession as to whether linear programming is suitable for industrial modelling, since economic techniques of production are not ordinarily linear. In economics, the field of the social sciences in which measurement has gone furthest, any predictions other than the short-term predictions of the movements of a known system have a very bad record. This is, of course, said in the preface to economic reports, but it is not said loudly enough – and it soon becomes easy for people to believe that they really know what is likely to happen. The tendency to indulge in abstract and unsupported postulations in United States strategic analysis is bad enough as is, with strong precedents in the political mood and context in which such analysis occurs, as well as in the methodology of several of its leading practitioners, and it would seem to be desirable that it receives as little further unsubstantiated input as possible.

To take another smaller example, in a paper on 'Grain Stocks and Protein Supplies' in the ORNL Annual Progress Report (1966–7) of its Civil Defense Research Group, there is a section on Soybean Production and Vulnerability of Processing Mills. The author makes a relatively happy discovery that on tallying United States soybean production by counties with the 9000-megaton mixed military industrial-population nuclear attack on the United States postulated in the book *Strategy for Survival* by Martin and Latham, 87 per cent of the soybean processing mills would survive, and hence be able to continue production. He discusses the feasibility of simple grinding methods that would re-route soybean flour for use in direct human consumption, the very low cost of presently importable machinery for this purpose, and the desirability of placing three such sets of machines in each present factory. The '87 per cent' figure of remaining mills is allowed to stand as a general indication of a post-attack soybean processing production capacity. How meaningful is this? No data is given on the power needs of such plants, their present sources, and the survivability of such generating systems and transmissions lines under the same Martin and Latham grid. Solvent storage at soybean mills is presently kept to a minimum, due to danger of its high inflammability (probably n-heptane, a petroleum distillation product). No data is given on refinery survival, on barrel or railroad tank-car survival which would be as necessary for the delivery of solvent as for the distribution of the finished product. In this case it was recommended that the soy flours be mixed with other grain flours for nutritional reasons. This assumes a like series of assurances – or questions – for the milling sequences of the other grains. Finally, how much present soybean land would be uncontaminated? how much seed would survive? and many other questions at the farming end – since soybeans are tractor cultivated and harvested, and would require fuel for tractors, etc. In short, when looking at the degradation in any one item, industry or capacity, these studies often or usually tend to ignore deficits in all the other interacting items, industries and capacities, and invariably come up with an optimistic result. Workers in this field of research have been heartened now that it is possible to introduce into their calculations reductions in damage estimates predicted by active (ABM) as well as passive (CD) ballistic missile defence systems.

In summary these criticisms imply, that despite the appropriateness of the general method of analysis in the NREC studies, the analysis does not support the predicted abilities to *reinstate* production because of the limitations in the data categories. Oddly enough, the PARM model appears to have now been withdrawn in favour of a more simplified model.

It is also pertinent here that there have been noticeable failings in the ability of the United States Civil Defense organization to deal with short-term, low casualty-producing domestic disasters. An example, which it was possible to watch develop in detail day by day, took place in St Charles, Missouri in 1967, when the rising of the Mississippi caused flooding in certain sections of the city. The rise of the river was anticipated, due to upstream rainstorms, and its exact peak crest, height and time estimated 2–3 weeks beforehand. Low points where flooding would occur were known. The means for dealing with the problem: partial civilian evacuation, constant patrolling of the dikes, construction of dikes, sandbagging, were well known. Here one had a single well-defined engineering problem, with a warning period and time to prepare for the emergency, yet the performance was publicly recognized as having been grossly inadequate, and there were subsequent recriminations at the state house. At times in the developing emergency the state Civil Defence director ostensibly had overall authority to deal with the situation. There have been other such recorded instances, a hurricane in Oregon in 1963 in which power lines went down and transportation was disrupted and in which the Civil Defence organization proved to be incapable of coping effectively with the situation. One might argue that these examples should not necessarily be taken as definite tests: perhaps the organizational network was poor, perhaps the director was timid, perhaps the state militia or federal troops were insufficiently utilized – in short, perhaps a better, or optimum, performance could have been turned in. Nevertheless, all these human inadequacies would obtain in the post-attack environments as well, while these examples took place in peacetime with all the organizational decision-making machinery and means to carry out such decision intact.

We might make some isolated remarks here on the civil defence role in certain other nations. The example of the CD programme in Great Britain is an interesting one. It was maintained all through the years for domestic political reasons, to still the political pressures of the territorial (home) army, as it was the only task that could be given to that organization to do, and not because it was presumed that civil defence efforts would be of much utility in the case of nuclear attack on Great Britain. In 1968 when economic pressures forced the British government to terminate the territorial army, it terminated much of its CD programme by the very same act. German civil defence expenditures for an air shelter system have been quite heavy, in terms of proportion of its annual defence budget, for many years. One should also point out that the oft-discussed Swedish civil defence efforts do not occur in the same context as those of a nation with a nuclear offensive striking force.

All of this previous discussion has dealt with goods and material, surviving stocks and production capacity. The second criticism is more crucial, and here one can group the PARM studies with all the others. Nearly every study the author has seen assumes, implies, and requires a tremendously rationalized post-attack world. It is hard to believe that human life, personal dispositions and attitudes, would be like that following a thermonuclear exchange of any serious proportions. Numerous examples from the reports could be quoted here to make the point, expressions to the effect that such and such 'should' transpire, or 'will' take place. Behaviour is always ordered, relevant and therapeutic. The studies carried out concerning social organization or disorganization under stress that have been carried out by the Disaster Research Group of the

National Academy of Sciences look at situations such as earthquakes, tornados, air crashes, and are orders of magnitude away from the effects, geographical ranges encompassed, and casualties that would ensue in a post-thermonuclear situation. In the former cases there has always been an 'outside' to supply food, medicine, psychological comfort and reconstructive potential, or if necessary to simply go about its business after a small sub-unit of itself might have been eradicated. These studies simply don't suffice; in fact, it is hard to believe that they are even relevant. A post-nuclear, or post-thermonuclear exchange situation in a major industrialized country is so wholly out of the range of previous experience that it is absolutely impossible, by any form of research, to produce any certain answers about the consequences.

It is probable that books dealing with two specific World War II phenomena, the concentration camps and Hiroshima, are more illustrative of the psychological pressures and human behaviour that might ensue. Examples of such books are: *Human Behaviour in the Concentration Camp* by Dr Elie Cohen, and *Hiroshima*, by John Hersey. One attempt in fact, to apply these lessons to CD and to the post-attack environment situation was made in the book and film by Peter Watkins, called *The War Game*.

Such rare and isolated OCD contracted reports as 'Vulnerabilities of Social Structure, Studies of the Social Dimensions of Nuclear Attack', by Human Sciences Research Inc. clearly indicate that satisfactory information about the *social* effects of nuclear attack are not known, and makes clear the degree of our ignorance in this area: 'There are many ways for society to fail to achieve recovery goals and only a few for it to succeed. It is problematical that it could succeed even with extensive prior preparations; it is unlikely that it could succeed without them.' To assert that social recovery is possible is to make a completely unwarranted assumption.

The requirements for an analogue to the PARM system, the description of a national social damage and social resources assessment system, requires fundamental extensions of only partially worked out social science techniques. It is not clear that all such requisites could be provided for. It is not clear that any sort of anticipatory planning or indoctrination could alleviate some of the psychological sequelae one can envisage resulting from mass destruction. The relationship of such psychological sequelae to the PARM type economic recovery programmes has never been raised. There does not seem to be a single study which deals with such psychological variables with any thoroughness or which even attempts to. These problems have the status of 'non-questions' in CD research; they just don't exist. It is not clear what the political or social cost of attempts to provide such extensive prior preparations would be to the combination of a Western democracy armed with strategic nuclear weapons. This is an element which does not necessarily enter into economic preparations or stockpiling. 'Social stockpiling' entails different social and political processes. What would be the effect of such programmes on the United States–Soviet Union Strategic confrontation via the tenor of the cold war? What would be the effects of such programmes on democratic societies? These last are of course the two most critical questions, aside from general questions of efficacy, that are omitted in the considerations of CD planning.

A society that was really prepared for nuclear warfare would be living in a state of siege. Some have speculated that public animosities towards a strategic opponent would be lower as a result of a massive urban CD programme. The understanding of this question cannot be arrived at by abstract logic or by wishful thinking. It is a matter of sociology; decision making processes of elites and of legislatures, and the mechanisms of imposing such decisions on a moderately educated electorate. The more modern human societies have been prepared for war, the closer they have been to war. What are these methods? Experience since 1900 has given us numerous instances at which to look for the answers. What are the implications of living in a state of seige? 'Systems analysis', which the United States strategic analysis makes much of, requires that when looking at one piece of a problem, one looks at all the other pieces with which it will interact. Whether an end likelihood, as a defence against which a particular proposal is suggested, is made more likely or less likely by the very proposal is surely the most pertinent related aspect. In the initiation, operation, maintenance and effects of that particular proposal – in this case civil defence – will it increase or decrease the chances for that end likelihood – which in this case is nuclear warfare – in a nation of 200 million, armed with nuclear weapons and a pervading domestic communications system? A careful analysis of domestic and international political consequences would show just how far the defensive preparation (CD), would go in producing the undesired end result; an example of the self-fulfilling prophecy. Yet these questions which are the basic, underlying constructs of civil defence programmes, are another group of 'non-questions' in CD research.

In 1964 the United States National Academy of Sciences National Research Council released 'Project Harbor Summary Report', a report by a summer study group on civil defence. The summer study had sixty-three 'participants' and eighty-eight 'briefers, consultants, observers'. The thousand-page, six-section, background report has never been released to the public. The six sections are entitled: Acceptance and Impact; Education and Training; Strategy and Tactics; Future Weapon and Weapon Effects; Immediate Survival; Post-attack Recovery. Few of the attendees brought with them positions sceptical of CD programmes, and there was consequently little adversary confrontation. Since there was subsequently serious criticism of the report in the scientific community, and among other things, questions on the degree to which the report reflected the material in the six-section reports, the Academy convened a second group in May 1967 to do an updating and revision of the original 27-page summary report, without any further great attention to new background material. There were twelve people in the group which met for two weeks. Only one person was invited in an 'adversary' role, and it cannot have been considered a very serious attempt to introduce reservations. As far as the sections dealing with 'Human Ecology' are concerned (pre-attack: Acceptance and Impact, Education and Training; post-attack: Post-attack Recovery) assumptions remained largely the same and the omissions largely the same as here described. The matter of post-attack recovery of natural ecosystems, plants and animals has not been discussed here. Here too there are large questions, but there is at least the possibility of doing some research, and though one has to make extrapolations, here at least one need not contend with the factors of human psychology nor the political requirements of a defence psychology in a peacetime democracy.

The expenditure that would be required for an urban blast shelter effort has always been a major United States political consideration. If anything, this will be a matter of even greater significance in the foreseeable future. Major Western nations are already showing severe strains over the diversion of resources from domestic needs to defence-related activities, both in terms of direct funding and of trained scientific manpower.

In conclusion, much has been written in the United States on the matter of the nuclear post-attack environment. Nearly all of it has been grossly optimistic, with a strong tendency to unrealism, in considering the economy bit by bit, in omitting a picture of the state of mind of the survivors and the resulting state of the social process, in omitting a consideration of the interaction of these two factors, and in omitting a consideration of the effects of an extensive CD shelter programme on the strategic nuclear confrontation.

# APPENDIX I

## Record of United States Federal Civil Defence Appropriations

(In Millions): 1951–1968

*The former Federal Civil Defense Administration*

| *Fiscal Year* | *Requested* | *Granted* |
|---|---|---|
| 1951 | $403·0 | $31·8 |
| 1952 | 535·0 | 75·3 |
| 1953 | 600·0 | 43·0 |
| 1954 | 150·0 | 46·5 |
| 1955 | 85·7 | 48·0 |
| 1956 | 75·4 | 68·7 |
| 1957 | 123·2 | 93·6 |
| 1958 | 130·0 | 39·3 |
| 1959 | 74·1 | 43·0 |

*The former Office of Civil and Defense Mobilization*

| *Fiscal Year* | *Requested* | *Granted* |
|---|---|---|
| 1960 | $101.7 | $52·9 |
| 1961 | 77·3 | 61·1 |
| 1962 | 104·5 | 86·6 |

*Office of Civil Defense, Department of Defense*

| *Fiscal year* | *Total requested* | *Amt. needing auth. legis.* | *Granted** |
|---|---|---|---|
| 1962 | $207·6 | (0) | $207·6 (49·6)† |
| 1963 | 695·0 | (495·0) | 113·0 |
| 1963-Sup. | 61·9 | (0) | 15·0 |
| 1964 | 346·9 | (195·0) | 111·6 |
| 1965 | 358·0 | (118·9) | 105·2 |
| 1966 | 193·9 | | 106·8 |
| 1967 | | | 101·1 |
| 1968 | | | 110·0 |

* No funds granted for programmes needing authorizing legislation due to lack of passage of such legislation.

† Transferred from OCDM appropriation to OCD/DOD.

## BREAKDOWN OF OFFICE OF CIVIL DEFENCE BUDGET REQUEST FISCAL YEAR 1966

| *Programme element* | *Budget request* |
|---|---|
| 1. Shelter Survey | \$36·3 |
| 2. Shelter Development | 3·0 |
| 3. Shelter in Federal Buildings | 7·8 |
| 4. Shelter Provisions | 52·6 |
| 5. Warning | 1·3 |
| 6. Emergency Operations | 13·3 |
| 7. Financial Assistance to States | 30·5 |
| 8. Research and Development | 15·0 |
| 9. Management | 14·6 |
| 10. Public Information | 4·0 |
| 11. Training and Education | 15·5 |
| (1966) TOTAL | \$193·9 |

# APPENDIX II

## Remarks Concerning Civil Defense made during Hearing before the Committee on Armed Services of the United States[2]

### G. CIVIL DEFENSE

The Civil Defense program proposed for FY 1969 contemplates no important change in basic objectives from those which I discussed last year. However, we have held the FY 1969 programme to the lowest possible sustaining rate, pending the end of the Vietnam conflict.

The major objective of the Civil Defense programme since 1961 has been the establishment of a comprehensive nationwide shelter system to help protect our population from radiological fall-out in the event of a nuclear attack. Most of this shelter is inherent in existing buildings but needs to be identified, marked and stocked with survival supplies before it can be considered truly useful. By the end of the current fiscal year we expect to have identified about 170 million spaces with a standard protection factor of 40 or more, of which about 101 million will have been marked and 55 million stocked with an average 14 days of supplies. Total shelter capacity should continue to grow in the future as a result of the continuing survey and design assistance efforts being conducted as part of the Civil Defense programme. In total, we can probably expect an additional 55 million spaces from these sources over the next five years.

A financial summary of the Civil Defense program, for which $77·3 million is requested for FY 1969, is provided on Table 2 (not printed in the published testimony). p. 129

*Question* 303: The testimony states that $76,800,000 is being requested for Civil Defense. For years some have felt that the military policy of 'gradualism' we have followed in Vietnam, has over the years, permitted other countries, primarily the Soviet Union to equip North Vietnam with the most sophisticated defense weaponry ever known. The chief justification for this policy of 'gradualism' – a policy that has cost so much in lives and treasure – would seem to be apprehension about doing something which would bring us into a nuclear exchange. It is clear from many reports that even a limited civilian defence setup would save tens of millions of American lives in a war. Is it now somewhat paradoxical, therefore, that as the United States continues to escalate the action of this war, we nevertheless are asking for less than 1 per cent of the military budget for civil defense?

*Answer:* As the question correctly implies, all of our studies show that an effective nation-wide fall-out shelter system would be, relatively, a highly efficient way of saving lives in a nuclear exchange. It was precisely because of this fact that the civil defense program was moved to the Department of Defense in 1961 and given new impetus. Since that time some 170 million spaces with a protection factor of 40 or more have been identified, of which 101 million have been marked and about 55 million stocked with an average 14 days of supplies. Ongoing programmes are expected to add some 55 million additional spaces over the next five years. In other words, we already have in being or under way a very large civil defense effort.

It is true, however, that even this amount of shelter will not provide fall-out protection for all of the population at all times (because of geographical maldistribution of the shelters *vis à vis* the population and the need to have redundant shelter capacity to accommodate both the daytime and nighttime locations of the population). In the early 1960s, the Administration proposed to stimulate the creation of the needed shelter spaces through a programme of Federal subsidies designed to maximize the fall-out protection potential of new construction. However, this proposal was not favourably received by the Congress. Since then the Defense Department has concentrated its efforts on the regular shelter survey programmes, including small and large structures and family homes.

As indicated in the Secretary's statement, the level of civil defense funding projected for FY 1969 does, in fact, reflect an ordering of priorities which calls for the program to be held 'to the lowest possible sustaining rate, pending the end of the Vietnam conflict.' Considering the immediate nature of Vietnam requirements and what it is hoped is the relative unlikelihood of an all-out nuclear exchange, the priority accorded the former should not seem paradoxical.

*References*

1 Remarks concerning civil defence posture appearing in Authorization for Military Procurement, Research and Development, Fiscal Year 1969, and Reserve Strength. Hearings before the Committee on Armed Services, United States Senate, Ninetieth Congress, Second Session, on S. 3293. Feb. 1, 2, 5, 7, 15, 16, 20, 21, 27, 28, 29; March 4, 14, 1968, page 341. U.S. Government Printing Office, Washington, 1968.

2 Remarks concerning civil defence posture appearing in Authorization for Military Procurement, Research and Development, Fiscal Year 1969, and Reserve Strength. Hearings before the Committee on Armed Services, United States Senate, Ninetieth Congress, Second Session, on S. 3293. Feb. 1, 2, 5, 7, 15, 16, 20, 21, 27, 28, 29; March 4, 14, 1968, pages 129 and 409. U.S. Government Printing Office, Washington, 1968.

M. NITA

# 2: Some Observations on the Consequences of Anti-Ballistic Missile Deployment

Contemporary civilization bears the imprint of a considerable scientific and technical revolution; we live at a time when scientific discoveries are taking place at a dizzy pace. Science has today become one of the decisive factors in the progress of all people, thanks to its large-scale and immediate application to the production of material goods. Humanity sees in the application of the successes of science and contemporary technology, which is the fruit of the indefatigable work of scientists everywhere, an important element in achieving its aspirations towards the development of civilization and a better life.

Unfortunately, however, there are at the same time tendencies existing for the utilization of scientific conquests and techniques towards destructive ends. The development of ABMs, for example, has important negative effects on some aspects of international affairs.

The analysis of the implications of the decision to introduce ABMs into the arsenal of the current means of combat must take into account in the first place the consequences of this action on the arms race. Indeed, if we analyse the long-term process of the development of armaments, it is easy to consider that the ABM system – despite the fact that it is far from having attained the necessary degree of perfection – can nevertheless generate a more perfected offensive system. This will in its turn engender another defensive system, and so on. It is for this reason that the decision to deploy ABM systems would involve contemporary society in a new phase of the arms race and it would augment the danger of the outbreak of a destructive nuclear war. This aspect represents, in the author's opinion, the source of all the unfavourable implications for contemporary international affairs which have been the subject of the papers presented in this book.

The struggle of all humanity in a series of persevering efforts designed to ameliorate the international climate and to set up the necessary premises for the strengthening of peace and international security is very likely to be seriously prejudiced by the arms race of which the development of ABM systems is a part. We are also turning our attention to the consequences which the continuation of the development of the ABM systems could have. The necessity of evaluating the efficiency of ABM systems obviously demands the undertaking of tests and trials in conditions which are as near as possible to reality. This could, at a certain moment, imply testing with nuclear warheads. This would contravene the steps already taken on the road to the stopping of nuclear tests.

Rumania, along with the socialist countries, and with all countries which desire peace, is fighting for the realization of general disarmament. At the same time, we are of the opinion that one can also undertake certain partial measures of disarmaments which lead towards the supreme goal of general disarmament. Many are aware of the actions taken by Rumania to ensure that the Treaty of non-proliferation is conceived as an integral part of a system of measures of which the end-product is nuclear disarmament and to offer equal guarantees of security to all countries, large and small, nuclear and non-nuclear. Rumania has insisted that this Treaty does not limit the peaceful use of nuclear energy, but, on the contrary, it assures limitless possibilities to all countries on the basis of equality and without any discrimination whatsoever for the utilization of the conquests of nuclear physics in a peaceful manner. As is well known, Rumania is a signatory of the Treaty. Unceasingly, Rumania has declared itself in favour of the conclusion of the Treaty for the non-proliferation of nuclear weapons. She has attempted with other peoples to make her contribution to its realization. Rumania considers this international instrument as an efficient counter to the atomic danger, as a measure of international *détente* and as a contribution to the reinforcement of peace and the security of peoples. This has, as a result, the development of confidence and collaboration between states and the opening of new perspectives for efforts directed towards nuclear disarmament. The declarations made by various states on the occasion of the signature of the Non-proliferation Treaty, which were concerned with the growing possibilities which are opening for the realization of new measures of disarmament attest to this fact.

The memorandum of the Soviet Government of 1 July enumerates a series of measures upon which the attention of negotiators in the domain of disarmament should be concentrated. Among these measures we must point out the intention to undertake discussions for the limitation and renunciation of the development of ABM systems of which the implications are the subject for analysis by the present symposium. Similar declarations have also been made by the United States' authorities.

The political effects of the development of ABM systems have been largely covered by other authors; one can easily deduce the negative economic effects which also stem from it. This is why we are limiting ourselves to the implications concerning the development of science in what follows as this is a problem closely tied to our professional activity. Two points should be made in this regard. The first is of a general nature and is concerned with scientific co-operation on a world scale. This has recently begun to acquire a considerable extension (to the benefit of all countries) and its future development is dependent on the continued amelioration of the climate of *détente*, on mutual confidence and on collaboration between all the countries of the world. The second implication is concerned with the possibilities of the development of space exploration for peaceful purposes. It is unanimously recognized today that the conquest of extra-terrestrial space is not only the result of scientific curiosity but that it also represents an imperative necessity for the progress of contemporary society. Since the beginning of the cosmic area, terrestrial utilizations of space techniques have been made, especially in the domain of tele-communications and meteorology. Other fields of space applications are geology, agriculture and forestry, hydrology and oceanography, geodesy and cartography, navigation, etc. In future it is anticipated that the area of space applications will continue to grow and that it will end by covering the most varied aspects of practical human activity so that finally space technology will fulfil one of the most important roles in the development of the civilization of the entire world. This fact will not fail to constitute an immense attraction towards space activity for all peoples and this will make them equally interested in promoting and extending international cooperation with a view to benefiting under its different aspects, and notably in their national economy, from the results and the possibilities that this activity offers.

It is at this point that the noble ideas of all humanity are opposed to the current tendency towards an escalation of the arms race. Although the system which has already been announced for an ABM has not yet even been set up, there is already talk, ever more frequent, of completing, indeed of replacing, these systems by systems of anti-missile satellites, by the utilization of artificial satellites as a means of delivering nuclear warheads with the aim of the destruction of terrestrial and other objects – without speaking of space systems of reconnaissance and communication which are already in the course of deployment. Thus one can immediately see the danger presented by the orientation of conquests and the successes made by space scientists in the development and the perfectioning of these types of armaments to the detriment of systems for peaceful use. The perfecting of present ABM systems, even if we suppose that this technique remains in its announced limits, would, of itself, necessitate the taking away of important intellectual forces from the scientific ranks of the international circuit. If this was not the case, these forces would have been able to make a substantial contribution to the extension of space exploration for peaceful purposes. Secondly, the involvement of military interests in space matters can only serve to restrict considerably and inevitably the possibilities of international co-operation of which we have been speaking, and this would clearly constitute a brake on the way to general progress. It is certain that if we were to continue to analyse the consequences of the use of ABM systems we would be able to point to negative implications in many other directions and fields of activity.

Starting from a clear examination of this phenomenon we can also find a key to its solution. General disarmament – and particularly its principal component, nuclear disarmament – international security, peaceful coexistence, the right of each people to its independence and national sovereignty, all these objectives are pillars in the edifice of world peace. It is necessary to conceive of this as a whole from which none of these elements can be missing. It is certain that in the realization of the supreme goal, that of disarmament, even partial measures have an important role to play if they enhance international *détente* and the lessening and reduction of the danger of nuclear war. In a position of incontestable importance among these measures is to be found a total ban on nuclear tests, a ban on the use of nuclear weapons and the cessation of production and a liquidation of existing stocks of nuclear weapons. In this situation the development of ICBMs and/or orbital systems designed to carry nuclear warheads and the means to counter them (such as ABMs) become clearly nonsense. Reason demands that these partial measures of disarmament be given general attention. Their realization must be sought with patience and especially with perseverance.

The objective analysis of contemporary realities leads to the conclusion that the elimination of war as a means of resolving state conflicts constitutes the cardinal question which dominates the whole of the problem of the organization of world peace. The peoples and nations of the entire world are profoundly interested in their economic and social development and the creation of standards of living which are worthy of the advances made by science and technology in our time. The essential premise for improving the lot of mankind to higher degrees of civilization and progress is the reinforcement of peace and international security. It is only in such a framework that the impetuous contemporary development of science and technology will be able to be fully developed for the good of all countries.

At present when the existence and continual perfection of the most destructive arms such as nuclear weapons and all the weapons systems tied to them creates dangers of an unprecedented gravity for humanity, the achievement of disarmament is the highest imperative. General disarmament and especially its principal element, nuclear disarmament, are fully in accordance with the necessity to assure equal conditions of peace and security to all countries and to permit the peoples to consecrate their efforts and their means to the work of construction and peaceful development.

Thus the Socialist Republic of Rumania is acting in a decisive manner for the realization of nuclear disarmament, for the liquidation of atomic arms and existing stocks and for the cessation of production of similar devices as this is the surest way to eliminate definitively the danger of nuclear war.

F. CALOGERO

# 3: The Arms Race and Public Opinion: A Suggestion

> There is a sort of mad momentum intrinsic to the development of all new nuclear weapons – R. McNamara (speech in San Francisco, September 1967).

The arms race is fuelled by several factors. One of these is the insufficient awareness, in public opinion at large, of the dangers associated with it. In fact, public opinion generally provides pressure towards an acceleration of the arms race rather than towards restraint. This is because the strategic and political arguments in favour of restraint are too sophisticated to gain widespread acceptance, in contrast to the primitive feeling that the possession of a larger weapon capability inevitably implies an increase in security. This situation is dangerous and should be recognized as such by all reasonable people, in particular by decision makers. Indeed, decision makers often appear to be forced, by this irrational attitude, to move in a direction which they themselves consider detrimental to the overall security of their own country and of the world. The decisions to deploy ABMs, first in the Soviet Union and then in the United States, provide examples of this phenomenon.

Decision makers should undertake, as an important priority, the task of educating public opinion in the widest sense, especially within their own country but also throughout the world, about the dangers inherent in the arms race. They should undertake this task according to a specially designed policy, and they should be prepared, in some cases, to sacrifice other aims in favour of it. It is clearly important that flexibility in the decision-making process concerning the arms race should be regained so that decision makers avoid being forced into options they themselves consider detrimental to overall security.

This endeavour has already been undertaken by some leaders in the major nations at least, to the extent of making the effort to explain the facts to the people. J. F. Kennedy and R. McNamara are prominent examples in the United States; N. Khrushchev might have played a similar role in the Soviet Union. But speeches are hardly sufficient to convey adequately the message to the public.

Many people believe that the present pace of development of technology is so much swifter than the rate of change of socio-ethical ideas that the latter cannot keep up with the former. Hence there is an imbalance between obsolete ideas on the one hand and new technological options on the other hand. It is argued that this imbalance cannot be redressed by any method short of the ultimate one, to which the imbalance itself will necessarily lead, namely a major catastrophe. This will have the double effect of stopping the progress of technology and remoulding the conscience of humanity by a sudden shock.

Others, less pessimistic, believe that before this final catastrophic exitus occurs a large-scale but less-than-total disaster might provide the shock necessary for humanity to embark on a drastically revised course, leading eventually to world government and disarmament. Such an event might result, for example, from the accidental explosion of one of the many nuclear weapons which are carried around the world in aeroplanes or which are loaded in the warheads of missiles. However, few, if any, would advocate such an event, let alone try to produce it.

The reason why speculations of this kind are usually only made in private rather than in public is probably because they might be taken as an indication that their proponent is not rational and in any case they do not lead to any reasonable suggestion regarding a specific course of action.

There is, however, one sensible, though marginal, line of action suggested by the above considerations, which is open to decision makers, rather than to private citizens, and which might be undertaken by them if they were convinced, not only of the need to mould public opinion towards appreciating the intrinsic dangers of the arms race, but also of the high priority of this goal. This consists of reversing the present policy of minimizing, or if possible suppressing altogether, all information concerning accidents connected with nuclear weapons or other weapons of mass destruction and, on the contrary, to provide as much publicity about them as is compatible with security requirements (interpreted in the narrowest sense). For instance, there is a rumour that an American plane found it necessary, some time ago, to drop a nuclear weapon in the megaton range over North Carolina, and that several of the safety devices, aimed at preventing the accidental nuclear explosion of the bomb, failed although at least one did work and the disaster was avoided. If this event really happened its impact on public opinion, in the United States and the world, could have been greatly increased if the details had been given maximum publicity. For example, the President of the United States might have personally announced the incident on television, adding a warning about the dangers implicit in the arms race. In effect, assuming this event really happened, there was practically no impact on public opinion, because all information was deliberately suppressed or effectively muffled.

The motivations which cause a government to attempt to suppress any publicity about accidents involving nuclear or other weapons are understandable. Among these are: security needs; adverse internal political effects (no government likes to be associated with failures of any kind); the propaganda opportunities offered to opponents abroad (for instance, the Palomares accident was used repeatedly by communist and non-aligned spokesmen to attack the United States); and the diplomatic and political consequences whenever the accident involves other nations (the Palomares and Greenland accidents are pertinent examples). These reasons explain the policy which has been followed up to now, and provide strong arguments for its continuation. They should, however, be carefully weighed against the arguments presented above, which point to an advantage of maximum publicity that is certainly much less tangible, especially in the short run, but possibly of more profound consequences.

It would, of course, be easier to take the action suggested if it were possible, presumably by some sort of tacit agreement, to initiate it in countries belonging both to the Western and to the Communist groups or, at least, to obtain agreement that restraint should be exercised by each side in exploiting, for propaganda and political purposes, the revelation of an accident by the other side. This would decrease the strength of the arguments – which would certainly be used by large segments of the establishment within each country – against the policy suggested here.

These considerations have been based on the hypothesis that to induce, in public opinion, dislike and fear of weapons of mass destruction is a useful, perhaps essential, step towards the development of a more sane policy. This hypothesis can be supported even though a similar policy has been used by those who were instead pressing for an increase in the level of armaments (again the ABM case appears appropriate; another example was the outcry about 'gaps' in various sectors of the arms race). The difference between the two uses of the same technique is that in one the policy aims at scaring the public by emphasizing the dangers inherent in the weapons of the enemy, whereas the other aims at scaring the public by emphasizing the dangers inherent in all weapons, and particularly one's own. The former policy has been pursued up to now; it has had the positive effect of making a policy characterized by caution acceptable, and indeed relatively popular; it has had the negative effect of making it more difficult to resist those who advocate an increase in the level of one's own armaments. The latter policy, which is the one suggested here, is quite different. Its purpose is to induce in public opinion an antipathy towards the new weapons of mass destruction, even one's own, with the purpose of providing the optimum conditions for the acceptance of more sophisticated arguments favouring restraint. This policy, apart from the price which would have to be paid to implement it, can hardly be expected to please all sectors of the establishment in any country; there would certainly be some, especially among the military, who would be very much against it, because they would fear that, if such a policy were effectively implemented, public opinion might be stampeded towards unilateral disarmament or non-armament. It is submitted, however, that this is not an imminent danger and that a lack of flexibility of this kind is improbable. On the other hand, the type of inflexibility described previously provides a major, although not an apparent, hindrance to any policy of reasonable restraint.

Finally, one other point should be mentioned. To suppress to the maximum extent possible all information concerning accidents associated with nuclear or other weapons is clearly in conflict with the right of the people to have access to knowledge of facts which directly affect them, a right that is basic to any political system which, whatever its institutional structures, claims to be 'democratic'. This argument might be taken up as a political issue. In this connection, it may be useful to re-emphasize that the policy advocated here is that such information be advertised, rather than suppressed, *to the maximum extent possible*, the proviso including all genuine security restrictions. It is unlikely that this is the policy now prevalent in any one of the countries involved in the arms race.

# Part III

## SUMMARY OF DISCUSSIONS

The following is a description of the discussions which took place at the Symposium. A significant fraction of the time was spent discussing the arguments put forward by authors in their papers; these arguments are not given in detail in the following summary but, instead, the reader is referred to the actual papers. No view is attributed to any one participant by name although minority opinions, when expressed, are given.

# Part III

# SUMMARY OF DISCUSSIONS

C. F. BARNABY; A. BOSERUP

# The Implications of the Deployment of Anti-Ballistic Missile Systems

## *Technical Aspects*

A few new points not covered in the earlier chapters, emerged during the discussion regarding the technical aspects of ABM systems. For example, it was explained that in the United States, although both Sprint and Spartan missiles could be used to protect ICBM sites, a final decision on whether to deploy even Sprint missiles around Minuteman ICBM sites has not yet been made and, in fact, money has not been appropriated in this year's (1968) budget for this purpose. Sprints will probably be used just to defend the radar system. Sprints could also, however, be used against missiles on fractional or other unusual orbits except those on the lowest energy orbits. The total cost of the Sentinel system, as planned, was given as about $5 billion, of which about 900 million is to be made available this year. Of this, 300 million is to be used for research and development, 400 million to begin production and deployment and 200 million to purchase real estate for sites for Spartan missile deployment.

The Sentinel system will have an operational capability by 1972 and will be fully operational by 1974. The Sentinel system based on Spartan missiles, will be ineffective against a sophisticated (e.g. Soviet) attack. Penetration aids like chaff and decoys could be made successful against it relatively easily. If, for example, a cloud of warheads and decoys were not too widely separated then both would be destroyed by Spartan missiles. The system, however, would be ineffective if the decoys and warheads were widely separated. Calculations on the effectiveness of Sentinel were necessarily largely hypothetical because some of the components of the system have not been tested to any extent. For example, although all the important parts of the Spartan missile have been tested to some degree no part has been thoroughly tested. The Sprint missile has been more satisfactorily tested but the characteristics of even this missile are not well enough known to enable a realistic assessment to be made of its capabilities under operational conditions.

It was pointed out that the American decision to proceed with Sentinel was not irreversible, and that it is still possible to stop and even reverse the deployment of ABMs. It was also noted that a large-scale civil defence programme is not, at present, associated with the Sentinel system. A relatively small amount of money, $77 million, has been reserved in the U.S. budget for civil defence programmes, but this will be used for such comparatively trivial purposes as providing primitive medical shelters.

The physical effects of nuclear explosives in the high atmosphere and space were discussed. It was concluded that megaton explosions from Spartan-type ABMs in space would cause negligible harm in terms of blast, shock, heat and radioactivity on the ground. The intense light flash may, in case of clear weather, cause retinal burns and flash-blindness of various durations in individuals. Nuclear explosions from Sprint-type ABMs would not cause serious effects on the ground because of their low explosion yields. However, the use of Sprint-type missiles may cause the opponent to maximize his warhead's megatonnage and to arrange that his offensive missiles will explode above the range of ABMs. Therefore, Sprint missiles cannot fully protect a population. In addition, a terminal ABM defence, even if deployed as a protection for all populated areas, would not make a civil defence effort unnecessary, since an opponent will always have the possibility of directing his offensive missiles against areas outside the populated ones and against off-shore targets, thereby causing casualties by radio-active fall-out.

### *The Effectiveness of ABM Systems*

The assessments of the effectiveness of ABMs vary from the very pessimistic, that they are virtually useless, to the optimistic assessments made by the U.S. military establishment. It was felt that for discussions of strategic issues the middle view of this range was reasonable. It was agreed that it was reasonable to claim that any single incoming ICBM can be destroyed by a single interceptor if the radar system is working effectively. Whether or not a number of warheads coming together can be destroyed depends upon several factors about which there is much uncertainty. It was stressed that, since there are no realistic operational data available, no true estimate of the effectiveness is possible. This uncertainty is likely to cause each superpower to overestimate the effectiveness of the opponent's ABM and, therefore, to overact to this deployment and, in addition, to underestimate the effectiveness of his own ABM system and consequently tend to deploy a heavier system than might be necessary for a given defence posture.

Some participants believed that the Sentinel system would be so ineffective that the first, primitive Chinese ICBM deployment would defeat it. Others argued that it might defend the United States against the Chinese for several years, although all agreed that it would never be effective against a Soviet attack.

Parts of the American administration and military establishment have defended the deployment of Sentinel primarily on the grounds that it is the beginning of a deployment which might be expanded. The deployment can certainly be politically defended as a system which could be expanded rapidly because of the early start. On the other hand, the system could be obsolete in a short time and its deployment would then prevent the spending of time and money which would be better invested in the design and development of a better system. A parallel was drawn with air defence which is now practically worthless. When the original decision on air defence was taken the system chosen soon became obsolete and a better system could have been developed if it had been decided to wait. It is nevertheless agreed that it would be politically easier to expand the system later if a start is made now even if the decision to deploy Sentinel technically ties the United States to a particular kind of defence system. It was emphasized, however, that

Sentinel is in effect, a putting together of components designed for a large heavier defence system and not optimal for any purpose; it is, in fact, the best that could be done at this time. Many anti-missile defence systems have been proposed in the past decade and several have been developed quite far. There is, however, always the tendency for them to become obsolete before they are operational and the final choice of the system to be deployed is a delicate balance between technical feasibility and economics. American estimates of the cost-exchange ratio have reduced from between 3 and 20 a few years ago to between 1 and 10 now. If these figures can be relied upon they give a measure of how technology develops.

*Effect of ABM Deployment on Superpower Détente*

The effect of ABM deployment on the *détente* between the superpowers was discussed in terms of (1) present and planned ABM systems, (2) possible extensions of these systems, and (3) future weapon development and deployment, particularly missiles with several independently targetable warheads (MIRVs). In addition, the effect of future weapon development on disarmament was discussed. It was recognized that even the present and planned limited deployment of ABM systems has had some accelerating influence on the arms race, even if it may not have significantly affected the *détente* between the superpowers. For example, the present level of Soviet ABM deployment has significantly influenced American decisions to increase the levels of nuclear weapons in the United States. In particular, it is probable that the decision to develop MIRVs was taken in response to the Soviet ABM deployment.

The development and deployment of MIRVs was seen to be particularly destabilizing. If these weapons are, in the future, developed to a high accuracy by both sides, and deployed with reasonably accurate ABMs, step-by-step disarmament will be particularly difficult. In any case, the deployment of MIRVs will add instability to the present balance in the area of strategic weapons. Moreover, any increase in military expenditure enhances the importance of the military sector in the decision-making process, and this can result in an adverse effect on the *détente* between the superpowers. All of these effects will, of course, be more severe if the present and planned ABM systems are extended in the future.

It was emphasized that the strongest arguments against the deployment of ABM systems are their effect on the *détente* and on the practicability of disarmament. It was pointed out that we are, at present, at a threshold in the arms race. The present situation is relatively stable and disarmament measures are practicable. However, new weapons development could easily lead to a situation from which disarmament would not be readily negotiable. It was realized that the invulnerability of nuclear submarines helped to stabilize the situation produced by the development of MIRVs and, therefore, the importance of submarines and other mobile launching platforms will increase as MIRVs become more accurate.

Submarines are, however, an order of magnitude more expensive to procure and maintain than are ground sites. The stabilizing effect of submarines would be considerably lessened by the deployment of effective ABMs. The serious consequences of future new weapon developments are, therefore, not significantly affected by the use of submarines and other invulnerable launching sites for ICBMs. For these reasons, the seriousness of the present situation and the great importance of future decisions on weapon deployment were particularly emphasized. Concern was also expressed about the adverse effect of ABM deployment on the consolidation and duration of the Non-proliferation Treaty and on efforts to reach a comprehensive test ban treaty. The proposed Soviet–American discussions on disarmament questions were, therefore, seen to be of extreme importance, and their success at the earliest date was regarded as crucial. The feeling was, however, that the forthcoming talks would be facilitated if the parties immediately stopped further procurement of ABMs and offensive missiles. Three specific measures, possibly suitable for early implementation by agreement, by unilateral decision or by mutual example were discussed. These were: restraint in ABM deployment; restraint in the development and deployment of offensive missiles; and a freeze in numbers of easily observable objects related to ABMs and offensive missiles, such as submarines and launching sites.

*Implications of ABM Deployment for the Arms Race*

There was general agreement that superpower deployment of ABMs would be a catalyst to the arms race, and this for several reasons: one is the obvious fact that the superpowers themselves will develop offensive techniques tending to offset whatever security gains the opponent might have hoped to achieve through ABMs. In particular all agreed that ABM deployment would lead to an increase in the number of offensive weapons and the claim by General Wheeler of the United States that the cost to the Soviet Union of deploying an adequate ABM system would be so high that the offensive weapons programme must be cut, was dismissed as ludicrous.

The deployment of MIRVs is one instance of this offsetting reaction. The tendency always to base policy decisions on conservative estimates accelerates the arms race and the more so the more uncertain the strategic configuration. Similarly, when deciding levels of offensive strategic forces the effectiveness of the opponent's defences will generally be overestimated. It was pointed out that even if only one of the superpowers engages in research, development and deployment of ABMs 'internal' arms races are likely to develop between the departments working on defence and on offence. Faulty assessment of the efforts of the opponent will contribute to an acceleration of the arms race on the missile-gap model. The fact that American MIRV development – supposedly intended as a response to Soviet ABM deployment – actually preceded the installation of a Soviet ABM system, was quoted as an example of this.

Many discussants felt that the arms race was now at a turning point and that in a few years time, depending on whether the superpowers decide to proceed with ABM deployment or not, there would be entirely different overall strategic configurations particularly if the present limited systems were to be expanded. Though it is not possible to spell out these configurations in detail, most agreed that ABM deployment, the unavoidable counter-measures and the consequent reorientation of the arms race would probably result in less, rather than more, stability as far as the relation between the superpowers is concerned. One major reason for this is the diminished second-strike effectiveness of missile-carrying submarines. Conceivably, ABMs may increase stability at the nuclear level between the superpowers and other nuclear powers, but this applies only when nuclear war is considered in abstract isolation and underlying tensions are ignored. The lesser vulnerability of the superpowers resulting from their deployment of ABMs may make them less willing to engage in constructive conflict settlement and thus have a negative effect on the international climate in general. This was particularly emphasized in relation to Europe and the Far East.

In judging the effects of ABM deployment on the arms race and on the international climate it was emphasized that this cannot be discussed only in the abstract terms of strategic theory, but that, in addition, internal implications must be taken into account. In the confrontation between hawks and doves ABMs will weaken the position of the doves and enhance the importance of the military sector in society in the superpowers as well as in other countries.

Coming at the same time as the signature of the Non-proliferation Treaty which explicitly requires of the nuclear powers that they make sincere efforts towards reduction of their arsenals, it was thought that the American decision to deploy ABMs was particularly detrimental and would gravely undermine the belief in the sincerity of the United States in this respect. The possibility that some non-nuclear powers might withhold their ratification of the Non-proliferation Treaty in order to put pressure on the superpowers towards discontinuing ABM development was discussed. It was thought that whatever the motives, a refusal by some countries to sign the Treaty would be unhappy as it might provide others with a cause (or an excuse) for doing the same and weaken the position of disarmament and arms limitation proponents in both camps. Instead, pressure on the superpowers might be exercised in other ways, for example by having a club of 'near-nuclear' countries meet from time to time and take stock of the progress in disarmament achieved by the superpowers.

It was emphasized that this is the first time that a formal declaration of intention to disarm has been signed by the superpowers and that they were clearly aware that it was in their own interest to make the Non-proliferation Treaty work. For this reason conditions may be particularly suitable in the next few years for advances in the field of nuclear disarmament. Other recent events make one optimistic in this respect, particularly the progress in the direction of an agreement on banning underground testing, and the United States acceptance of the Soviet proposal to discuss a freeze and cutdown of offensive and defensive arms. It was proposed that Pugwash should discuss such problems as could render support to these negotiations and make them more successful.

Finally, some aspects of Mr Mortensen's paper (page 108) relating to the implications of ABM deployment for the arms race were taken up. Serious doubts were raised concerning the significance one could attach to the figures for the cost-exchange ratios of offensive and defensive weapons as used in that paper. Their values are derived from abstract and highly simplified models based on rather arbitrary assumptions. Many such models can be devised, all giving different figures, and they, therefore, mostly serve as rationalizations. It was also thought that since reasons of domestic policy and many other motives which bear no relation to considerations of military strategy play an essential part in major decisions on weapons policies, cost-exchange ratios and other concepts derived from strategic models, even if accurately known, may have little impact on the way the arms race develops.

In judging from a broader point of view the desirability of an ABM system, the central issue, it was believed, is not the level of armaments as such but the resulting stability. Most discussants thought that while one had now reached a level of relative stability between the superpowers compared with earlier situations, for instance when ICBMs were being deployed, ABM deployment would accelerate the rate of technological development in the field of strategic weaponry and once more make the situation more unstable and uncertain. In addition, ABMs are a measure of only temporary importance and the improved methods of attack they will generate may turn out to be of greater permanence.

*The Effect of ABM Deployment on the Test Ban Treaty and the Non-proliferation Treaty*

In the discussion of the relationship between ABM deployment and disarmament issues, the paper by Mr Leitenberg (page 56) on the test ban treaty and its relation to ballistic missile defence was briefly discussed with particular reference to the Test Ban and Non-proliferation Treaties.

It was generally felt and strongly emphasized as a counter to widespread and unfounded optimism that the Non-proliferation Treaty would not be viable for more than a few years unless substantial disarmament measures were taken by the nuclear powers, as otherwise the other signatories were bound to withdraw from the treaty one by one. By speeding up the arms race ABM deployment would greatly increase the prospect of this development. Furthermore, mere perseverance by the superpowers in ABM deployment would necessarily have an impact on the belief in their sincerity as regards their obligations under the Treaty.

Even apart from this, and to the extent that ABMs *per se* increase the offensive capability of the nation deploying them, the deployment itself is a violation of the spirit of the Non-proliferation Treaty. Finally, it was pointed out, with particular reference to Western Europe, that, because an ABM system must be able to react at extremely short notice, any system deployed in nations which are not now nuclear would violate the Treaty since under those conditions no effective control by the country applying the warheads would be possible.

The effect of ABM deployment on the viability of the Moscow Test Ban Treaty of 1963 and the prospects for extending it to include underground tests as well was also considered. In a recent statement John S. Foster, the director of Defense Research and Engineering of the United States Department of Defense, maintains that previous fears that the treaty would have to be broken in order to test the ABM system now in the process of deployment were unfounded as no above-ground tests would be required. This statement was viewed by the group with some scepticism. While separate components like rockets and data-handling equipment can be tested without violating the Treaty there would be at least a considerable temptation to test the mechanism whereby incoming missiles and warheads are destroyed and extrapolation from underground experiments, it was suggested, would provide rather limited guidance in this respect.

While perhaps testing in the atmosphere or in space is not strictly required for a limited system it was generally admitted that this would definitely not apply if the United States were to deploy a large system. Even with the present system, however, there would be political reasons in addition to the technical ones for submitting it to a full test. While complex weapons like these can perhaps work without complete testing, the credibility both at home and abroad, may be more problematic.

It was also pointed out that if ABM development is allowed to continue it is likely to undermine the Test Ban Treaty in other ways. It would enhance the importance of further development of warhead characteristics and the mutual suspicion that the opponent was more able than oneself to develop the weapons using underground tests only might create a strong pressure on both sides for breaking the Treaty. Even the signing of a complete ban including underground tests would not resolve this difficulty, as there is a limit below which underground explosions cannot be detected or identified. This situation of course applies equally to any other phase in the arms race where improvements in nuclear weapons technology are believed to be both feasible and potentially of great military significance. Thus the fact that ABM deployment leads to a technologically new phase in the arms race rather than simply a growth in stocks makes the continuation of the test ban problematic.

It was thought that intensive research and development in ABMs would be particularly detrimental to the completion of a treaty banning underground tests. In this connection it was brought up that the study group on seismic detection of underground explosions recently brought together by SIPRI had established that explosions down to 10 kilotons (in granite) could now be detected and identified without on-site inspections, and it was even claimed, that this lower limit was probably a conservative estimate. Hence the technical problems are now largely solved, and it was felt that it would be tragic if ABM deployment were to spoil the chances of a complete test ban treaty coming into effect.

*Effects of ABM Deployment on China and on other Asian Countries*

The implications of the fact that the American deployment of ABMs had been officially presented as directed against the People's Republic of China were considered on the basis of Mr Leitenberg's paper (page 139).

Most of the discussants – although not all of them – agreed with the author that the Chinese aggressiveness is probably largely a myth and that, if it is not, it is to a large degree the result of international isolation and of past and present containment policies. In consequence, the deployment of ABMs against China was seen as particularly unhappy because it would further exacerbate relations with this country and would reinforce the negative stereotypes on both sides. In this way the expectation of (verbal) agressiveness on the part of China might be self-fulfilling. It was claimed by some, but strongly challenged by others, that the Korean war provided a demonstration of Chinese aggressiveness.

It was generally held that hitherto China's behaviour on the international scene has been rational and predictable, and that in any foreseeable future, and certainly for as long as the present American ABM system will last, it is highly unrealistic to assume that China should want to launch an attack on the United States, whether ABMs are deployed or not. This was one of the reasons why many felt very sceptical of the official claims that the American ABM deployment is in fact dictated by a perceived military necessity or usefulness in relation to China. For the United States to need a defence against China there must be both a Chinese willingness to engage in military adventures and a capability for its doing so, for instance in the form of a conventional superiority. It was strongly emphasized that there is no evidence to suggest that there is either intention or capability. China's ability to exert nuclear blackmail on the United States seems remote both from the point of view of intentions and of capability and it appeared to the group that the only conceivable *military* motive for the United States to deploy an ABM system against China was to maintain an absolute superiority and hence the possibility for the United States itself to engage in nuclear blackmail. Opinions differed, however, as to whether even this type of blackmail is in fact possible or whether it is merely a far-fetched armchair theory. It was pointed out that during the Korean war the American nuclear superiority and invulnerability was much larger than it is ever likely to become again, and yet there was nothing to support the claims by President Eisenhower that American nuclear weapons had had any decisive effect on the war.

Probably the American decision to deploy ABMs had many different motives among which fear of the Chinese nuclear programme and even military reasons in general were not the foremost. Domestic political reasons may have been equally decisive. It was generally thought that at present developments in China's military programmes have little effect on defence decisions in the United States and, quite possibly, that the converse is also true. At any rate the group could see no clear technical connection between the Chinese and American programmes. Hence the American decision to deploy ABMs may have only secondary military effects in South-East Asia, but it seemed undisputable to most that it would have important, immediate and deplorable consequences for the political climate.

Most discussants felt that faced with a real or imagined Chinese nuclear threat it would be best if the superpowers were to rely upon deterrence instead of protection as they do among themselves. Probably China will not accept indefinitely the containment and encirclement she is now subjected to, and if the superpowers maintain an absolute superiority and political and military freedom of action in respect to China this will simply postpone her integration into the international community. Previous experience tends to suggest that the ease in tension between the United States and the Soviet Union came when both felt relatively secure. China is a potential superpower and, therefore, should be expected to adopt policies, ambitions and aims like those of the other superpowers. The fact that she is acquiring this status while facing internal problems provides additional uncertainty. On the other hand the People's Republic of China has made a solemn pledge not to use nuclear weapons first and is in fact the only nuclear power to have done so.

Views differed somewhat as to the effect American ABM employment might have on nuclear proliferation in other parts of Asia. On the one hand it was suggested that without ABMs the American guarantee might lose in credibility and hence Asian governments might wish to develop their own weapons. At the other extreme it was pointed out that ABM deployment was more likely to accelerate such a development. In the first place it would lend credibility to the Chinese weapons. Sooner or later one would have to admit that the ABM system no longer provided much protection for the United States. The Asian governments might then feel the need for a national deterrent. If the United States were to emphasize the adequacy of their present deterrent rather than opting for protective devices like the ABM such problems would be less likely to arise later. In addition, in judging the desirability of ABMs from an Asian point of view it should be kept in mind that in the Eighteen-Nation Disarmament Committee the Indians had been consistent in denouncing ABMs and one might do well in the West to take note of what the Asians say they want, not what Western theorists think they ought to want.

*Effect of ABM Deployment on Bipolar Stability*

The discussion of this topic and of the effect of ABM deployment on the credibility of the lesser of the present nuclear deterrents, centred around Dr Carlton's paper (page 126) which argues that for the relations between the superpowers massive retaliation should be replaced by the doctrine of limited strategic nuclear war while the lesser powers should be prevented from cataclysmic interference in the central balance by superpower deployment of ABMs. Both the premises and the conclusions of this paper were challenged during the discussions.

First, the concept of limited strategic nuclear war was considered to be merely a mental construction and not at all likely to work in practice. It presupposes so much by way of control, mutual agreement and communication between the superpowers that less suicidal and infinitely preferable agreements would sooner come within reach. Besides, the Soviet Union has consistently refused to accept this doctrine as a realistic possibility and this alone would seem to rule it out.

Also the alternative between limited strategic nuclear war and a doctrine of massive retaliation was considered to be neither realistic nor exhaustive. On the other hand it was pointed out that massive retaliation in cases where only limited objectives are involved has been abandoned and is, therefore, not a suitable yardstick by which to judge other strategic doctrines. It was stressed that there are other alternatives besides limited and total nuclear war, even where the so-called vital interests of the superpowers are involved, and it is only by artificially narrowing down the choice to these two that limited nuclear war becomes the lesser evil. Only the simplistic view that a partial defeat on one issue must eventually lead to surrender on all other points as well, makes war appear unavoidable when the *status quo* is disturbed. Any reasonable policy it was felt, must leave room for partial or temporary defeats, compromise solutions and so forth, and the example of the Cuban missile crisis demonstrates that it is of very limited importance whether a country loses in a particular political confrontation.

As everybody agreed that ABM deployment by the superpowers would to some extent accelerate the arms race between them, the central issue was whether deployment could be justified by the fact that it would reduce the likelihood of nuclear war being initiated by smaller powers. In this connection it was pointed out that in Dr Carlton's argument ABMs only have the function of defending the capitals of the superpowers against attack by third powers but since there are alternative command centres ABMs are in fact not a necessary element in the theory. Nor, evidently, are ABMs sufficient to prevent catalytic war if they only protect the capitals. In the argument of the paper, however, anything but a thin ABM system would upset the stability of the central balance. Altogether the paper seemed to provide too abstract an answer to an equally abstract danger since catalytic policies by lesser powers, being suicidal, seem highly unlikely, and the smaller powers would presumably be more interested in fighting their own battles than in interfering in the central balance. For example, China is presumably more interested in what goes on in Taiwan than in the United States.

Finally, the objection was raised that proposals like this one would at the very best provide a breathing space of some years during which the present bipolar structure of the world would be maintained. At the same time, however, it would freeze the *status quo* and, therefore, merely postpone the solution of political problems which give rise to the tension and hence to the danger of war. A sensible strategic doctrine, it was maintained, must leave sufficient flexibility in international relations to make possible the solution of conflicts and a doctrine which freezes relations between states into a fixed pattern is likely instead to magnify conflicts in the longer run. The argument of the paper was seen as one example of the danger there is in pursuing policies which are only advocated because the likelihood of a nuclear war occurring in the immediate future is considered to be very high. Such policies easily become strategies for conducting war rather than strategies for preventing it.

### *Consequences of ABM Deployment for Neutral Countries*

The consequences of ABM deployment for neutral countries in a crowded area like Europe were discussed. The main physical effects from neighbouring countries employing ABMs would be the hazards from the light flash of missiles exploding in space or in the upper atmosphere as well as the dangers resulting from stray missiles. On the other hand an effective ABM defence in a neighbouring superpower might reduce radiation hazards somewhat. It was strongly emphasized, however, that these and similar effects were trivial compared with the catastrophic consequences of a major war and that it would be irresponsible to suggest otherwise or to base policies on any other conception.

The possible military and security advantages for a neutral country itself deploying ABMs were also thought to be quite marginal while the security drawbacks may not be. Deflecting ballistic and anti-ballistic missiles off their course, so that they were steered away from third countries was considered to be technically unfeasible at present, though not necessarily inconceivable in the future. Again it was strongly felt that in deciding whether to build ABM defences or not, the one overriding security consideration for a neutral country is the effect this would have on the international climate and the likelihood of war. Improving the political climate would be a lot more important to a neutral country, especially in a place like Europe, than shooting down stray missiles or any similar protection measures one may conceive of.

## ABM DEPLOYMENT IN EUROPE

The possibility and implications of ABM deployment for Europe was discussed at great length, both from a technical and from a political point of view. While in technical and economic terms such a system was thought to be somewhat ahead in the future, political pressures it was felt, could build up rapidly and an ABM system, even if militarily pointless, would have grave and far-reaching political implications.

### 1. *Technical Aspects*

It was pointed out that from a technical angle a European ABM system would be an entirely different matter from that which is being considered for the superpowers. The area to be defended is much smaller and much more densely populated so the problem of radioactive contamination by exploding ABM warheads in the course of a major exchange would become prohibitive, even assuming that the system were completely effective in preventing penetration. Furthermore, the distance between the blocs is so small in Europe that it is no longer a clear ICBM situation as it is for the superpowers. A defence would have to be effective against low-flying missiles as well and possibly even against artillery shells. This would further increase the superiority of the offence over the defence, and decoys and non-ballistic trajectories might saturate or defeat the system rather easily.

ABM systems for protection of areas in Europe may range over a wide spectrum, between ordinary air defence systems and Sentinel-type systems. In fact, an effective system would have to cover the whole range. Warning times would vary from almost zero to about 20 minutes. The shorter the range of the incoming missile the shorter is the warning time and the smaller the area protected by a single ABM interceptor. Therefore, the cost of protection per unit area becomes increasingly greater. If a defence system were established against one type of short or intermediate range missile, a potential enemy would be likely to deploy another type of missile, thus making the defence system almost meaningless. A credible ABM defence in Europe (or in any place where two opponents are in close proximity) would, therefore, have to deal with all possible ranges of offensive missiles, which would make such a system prohibitively expensive. It was stressed, however, that a limited ABM capability against short range missiles could be bought for a limited cost, and this in turn may be the basis for pressures to acquire such limited capabilities, whether militarily meaningful or not.

Hence it seems that an ABM system in Europe can at best be effective as a point defence of offensive missile sites. (The radioactivity problem would make deployment of present-day ABM systems around cities an absurdity.) In this case, however, it would be both cheaper and more effective to place the missiles on ships or in submarines instead of protecting them on the ground.

It was thought that the problem in Western Europe as far as nuclear weapons are concerned, arises primarily from the geographical proliferation of the tactical nuclear weapons. Most observers agree that these weapons cannot be used without escalation proceeding to a level where Europe would be totally destroyed. ABM deployment in Europe – whatever the capability of the system in war – would therefore be counter-functional, implying that the use of small nuclear weapons is considered to be feasible. There was general agreement that today there is no rational need for nuclear weapons, including tactical nuclear weapons, in Western Europe, and that the area would be safer without them. Furthermore, present-day air-lifting capabilities provide a technological case for a complete withdrawal. Since Europe, whether East or West, cannot be defended with ABMs it was strongly emphasized that one must instead invest the effort in lowering the probability of attack.

In Europe the distinction between offensive and defensive missiles would be blurred, not only because an improvement in defences in itself represents an improved offensive posture (which applies to the superpowers as well), but also because the missiles and explosives of an ABM system might be used offensively over the short distances involved.

It was pointed out that an ABM deployment in a non-nuclear European country could not fail to cause the Non-proliferation Treaty to be violated, as warning times would be so short that foreign authority over the use of the nuclear explosives in the ABMs cannot be maintained. In fact, *any* ABM deployment in Europe would be in contradiction to the spirit and sense of the Non-proliferation Treaty and, if undertaken by a non-nuclear power, to its letter.

The possibility of a purely hypothetical anti-ballistic point defence in Europe using chemical explosives instead of nuclear was discussed. Such a system would probably have to rely upon actual collision with the incoming missile, and is hardly likely to become feasible for many years. It was generally considered that most of the technical case against a nuclear-based system would not apply for a non-nuclear one. However, the political case would still apply in part; its deployment might still be detrimental to the political climate and postpone arms-limitation agreements, and it might promote adventurism by providing an illusion of protection.

2. *Political Aspects*

Even though an effective ABM system in Europe would be technically unfeasible today and would be immediately countered by improved offences, and even though it would be suicidal to use such a system – if only because of the problem of radioactive contamination – the present situation was viewed with grave concern. The deployment of an ABM system in Western Europe has been discussed at various levels recently. In May (1968) the Nato Defence Planning Committee decided that present circumstances did not justify deployment in Europe, but recommended constant review of developments in this field. Even so, many participants in the Symposium expressed their apprehension that it might become extremely difficult to prevent the development of European ABMs at a later time.

The technical problems may come to look different in a few years' time, but even if they remain, there may be psychological and political reasons to proceed with such a system if the superpowers go along. If such is the case, it was thought, some weapons system for Europe will be invented sooner or later which, on paper, can be made to look feasible at least to some military. The further development is then difficult to stop. It seems to be of limited use to point out that the weapons will not work, as in Western Europe (and elsewhere too) there is a history of buying weapons that do not work. Besides, one is accustomed in Western Europe to systems which provide partial security at the very best, and if the main motivation is political, the military rationale superimposed on it is anyhow secondary, and even the cost argument may not be very important because token systems can be bought fairly cheaply. Also, the cost of large systems is usually underestimated by a factor of three in the first place, and sometimes by as much as a factor of ten.

It was stressed that even though on paper it may look differently, the cost-effectiveness ratio of an effective European ABM system would probably be essentially infinity because the system itself calls for counter-measures which are easily implemented and the cost of offsetting these counter-measures must be included in the first place. Some felt that similarly the cost-effectiveness of an American or Soviet system would also be close to infinity, at any rate in the somewhat longer run.

However, even this way of presenting the issue understates the case against a European system because when it is discussed in terms of monetary costs alone there is an implicit assumption that the effectiveness in terms of security is positive when in fact it may very well be negative. A meaningful assessment of the costs and benefits of an ABM system is much more intricate than a simple determination of its monetary cost and its efficiency in intercepting incoming missiles. The alternative is not between an arms race without ABM and the same arms race with an ABM system on top of it. Most probably entirely different strategic configurations would result under one or the other hypothesis. The development of MIRVs, the lesser overall stability and the lesser stabilizing effect of the missile-carrying submarines illustrate this point. In particular the effect on the political climate, and on the likelihood and scale of war must be taken into account. Seen in this way the penetration factor – unless very nearly zero – is largely immaterial.

Complicated military hardware may be bought as a way of yielding to, or in order to pre-empt, charges by the opposition at home that security is insufficiently assured, or it may be felt to be necessary in order to relieve frustrations and compensate disaffected military personnel, as an excuse for subsidizing the aerospace industry, or for various similar reasons. The situation as regards Britain was thought to be particularly problematic as the emergence of ABMs coincides in time with the retreat from East of Suez. In France, the development of the *Force de Frappe* is still going on and in evaluating an ABM system it was believed that technical and economic considerations and considerations of efficiency would play a greater role.

It was strongly emphasized that the attitude of the United States is crucial to the whole issue. Technically, politically, and perhaps economically as well, their assistance would be necessary, and so far all major weapons systems in Europe have required an American initiative. One danger, it was pointed out, is that one day the United States will see ABMs as important for promoting co-operation and unity in Europe, as used to be the case with the Multi-lateral Force (MLF). All reasons advanced in favour of (and against) MLF seem to be valid here as well, and on the basis of that experience it appears essential that the West Europeans be sufficiently aware of the negative effects of ABM deployment to be able to resist American pressures.

It was pointed out that the two major options concerning Western Germany, a denuclearization and subsequent reunification of Germany or its further integration into the Western Alliance, were closely related to the issue of ABMs in Western Europe. It was believed that if ABMs were taken up as a second MLF it would probably not be with European defence, but rather with the political implications in mind.

It was thought that reviving the idea of a nuclear-free zone in Europe would be much preferable, and that a complete withdrawal of nuclear weapons would improve European security considerably. It was mentioned that the political problems with such a proposal might be less than previously since reciprocity, while desirable, would not be strictly indispensable under present circumstances. The problem, it was pointed out, has so far been largely related to Germany, and the maintenance of the political *status quo* has apparently been a more dominant preoccupation than the maintenance of a strategic *status quo*. It was strongly emphasized that any independent ABM deployment in Europe would politically be in contradiction to the present attempts to reach an agreed settlement in Europe. All technical, financial and security reasons were of secondary importance compared with this political reason.

Another concrete proposal was discussed and met with support from all sides. This was the idea of pressing for commitments on non-first use of nuclear weapons. China has already made a declaration to this effect, and it was found it would be useful if the Europeans would support the Soviet proposals in this direction. Although the nuclear response to a non-nuclear attack is still official NATO policy, it was pointed out that according to the latest official United States sources the U.S. Defense Department does not any longer believe there is a conventional Soviet superiority. Hence the United States would not have to use these weapons first. Once non-first use commitments had been made, one could start asking why the tactical nuclear weapons were in Western Europe, and whether to remove them.

Some doubts were expressed as to whether Britain and France would join a non-first use declaration, but presumably it is of relatively minor importance whether they do. A declaration by the superpowers alone would have a salutary effect on the international climate as well as on military planning and thinking.

*Conclusions*

The consequences of rapid technological advances in all its aspects, including weapon technology, were regarded by the Symposium to be of fundamental importance for the future of International Society. The participants stressed that it was essential for world peace and security that the non-proliferation treaty should be successful and lasting, and that it should be followed by other measures, in particular that there should be a successful conclusion to the forthcoming disarmament talks between the Soviet Union and the United States. An optimistic feeling was, however, expressed that the conditions may be ripe for advance in nuclear disarmament. This was based on the fact of the signing of the Non-proliferation Treaty, on the progress in the direction of an agreement on banning underground testing, and on the American acceptance of the Soviet proposal to discuss a freeze and cut-back of offensive and defensive weapons. It was found that the time had come when a non-first use declaration by the nuclear powers might be within reach and that the Soviet proposal to this effect should be given full support. Such a declaration would have a salutary effect on the international political climate, on the problems relating to the stationing of tactical nuclear weapons in Europe, and on military planning in general.

Fears were expressed, however, that Man may not be able to control the rapid advances made in technology, and ABM deployment was cited as a glaring example of the instability in technology itself. Technology makes new weapons available; because they are available political pressures build up within states for their acquisition; political leaders succumb to this pressure and invent policies to justify the acquisition of the weapons. Once the weapons are acquired, strong pressures develop for their large-scale deployment, even in the absence of sensible strategic arguments.

There was a strong feeling among the participants that measures like the comprehensive test ban treaty and restrictions in using, stockpiling and producing nuclear weapons, which are being discussed in the context of the disarmament talks in Geneva, would dispose of the ABM problem and would be the best way of eliminating the danger of nuclear war. It was pointed out that whatever technical advances are made the offence will continue to be tactically superior to the defence, partly because its aims are more limited and more flexible, and partly because the offence has the initiative and, therefore, chooses the means. Technically, any ABM system is, therefore, fairly easy to defeat.

However, what is at issue now is much more than the technical advantages and drawbacks of ABMs as both are likely to be only temporary. The real issue is the possibility of nuclear arms control measures and of big-power restraint. It is the hopes of denuclearization and of disarmament measures versus a new spiral in the arms race with the additional strain it imposes on the international climate and the instability it adds to the strategic configuration. It was pointed out that a defensive system is, for political reasons, more difficult to dismantle than is an offensive system, because the case for a protective measure can be argued in simple moral terms, whereas the case against it requires more sophisticated arguments. Also, wishful thinking tends to perpetuate the belief in the utility of defensive systems. This is an additional reason why nuclear disarmament will become comparatively more difficult if ABM systems are allowed to proliferate. It was emphasized that the decision to deploy or not to deploy an ABM system is not likely to be dictated by military and technical considerations but by political ones, although, of course, a strategic rationalization would in most cases be provided.

Yet, if the causes for ABM deployment may bear little relevance to strategy this is not so for the effects. An ABM system, or any other protective system, will make it more tempting to increase the offence – whether in addition it increases the temptation to strike first depends upon the rate of improvement of the defence over the offence. It will increase the role of the military in decision-making and increase public suspicion in both blocs. Furthermore, the prospects for disarmament would be much worsened by the development and deployment of accurate MIRVs, particularly with ABM systems. If both sides acquired these systems, step-by-step disarmament would be extremely difficult, since neither side could move from its position without passing through a phase of extreme instability.

# APPENDIX

## Extracts and Conclusions from Two Official American Statements on ABM Deployment

As part of a statement to a joint session of the Senate Armed Services Committee and the Senate Sub-Committee on Department of Defense Appropriation in January 1967, Mr McNamara, then the United States Secretary of Defense, gave a detailed analysis of the advantages and disadvantages of the United States deploying an ABM system against a possible Soviet nuclear missile threat. Because of the importance of these statements in any discussions on ABM systems the principal conclusions drawn from Mr McNamara's estimate are now given in some detail.

### (*a*) MCNAMARA STATEMENT TO SENATE COMMITTEE

*On the advisability of deploying the Nike-X System*

The Soviets have it within their technical and economic capacity to offset any further 'damage-limiting' measures we might undertake, provided they are determined to maintain their deterrent against us. It is the virtual certainty that the Soviets will act to maintain their deterrent which casts such grave doubts on the advisability of our deploying the Nike-X system for the protection of our cities against the kind of heavy, sophisticated missile attack they could launch in the 1970s. In all probability, all we would accomplish would be to increase greatly both their defence expenditures and ours without any gain in real security to either side.

*Soviet ABM System*

It must now be assumed that by the early 1970s some 'reasonably effective' form of ABM defence would be deployed round all the major Soviet cities, using Galosh missiles – already being employed to surround Moscow – or a combination of Galosh and other missiles, at a cost of at least $20,000,000,000 to $25,000,000,000.

*Anti-Chinese United States ABM Systems*

Despite the firing of a nuclear armed missile over a few hundred miles in October 1966, it appeared unlikely that the Chinese Communists could deploy a significant number of operational intercontinental ballistic missiles (ICBMs) before the mid-1970s. Moreover, as the lead-time required for China to develop such a force was greater than that required for the deployment of the United States defence against it, the Chinese threat in itself would not dictate the production of an American ABM system at present. An 'austere' ABM defence costing about $3,500,000,000 inclusive of nuclear warheads could probably preclude damage from Chinese missiles almost entirely in the 1960s, and relatively modest additional outlays could 'probably limit the Chinese damage potential to low levels well beyond 1985'.

*United States Strategic Forces*

In order to counteract the deployment of ABM systems round the principal Soviet cities, it was proposed in the United States programme for fiscal year 1968 (i.e. 1 July 1967 to 30 June 1968) (*a*) to produce and deploy the Poseidon missile; (*b*) to produce and deploy improved missile penetration aids; (*c*) to increase the proportion of Minuteman-3 missiles in the planned force and provide them with an improved third stage; and (*d*) to initiate the development of new re-entry vehicles specifically designed for use against targets heavily defended with ABMs. The net effect of these actions would be to 'increase greatly the overall effectiveness of our assured destruction force against the Soviet Union by mid-1972'. Even if the Moscow-type ABM defence were deployed at other cities as well, the proposed United States missile force alone could inflict about 35 per cent fatalities – 86,000,000 – on the Soviet Union in 1972 after absorbing a surprise attack.

As regards Communist China, a relatively small number of warheads detonated over fifty cities would destroy half of that country's urban population and more than one-half of her industry, so that the forces planned for fiscal years 1968–72 'would by themselves give us an assured destruction capability against both the Soviet Union and Red China simultaneously'.

The deployment of a substantial Soviet ICBM force with hard-target kill capacity might pose a threat to the Minuteman missiles, while an extensive and effective ABM system might then be able to intercept and destroy a significant portion of America's residual missile warheads, including those carried by submarine-launched missiles. The development and production of the Poseidon missile had been authorized so as to hedge against the possibility of such a threat to land-based missile forces, while the development and deployment of a new advanced ICBM was also under consideration. At the same time the deployment of the Nike-X ABM system as a defence for the Minuteman force would offer a 'partial substitute' for the possible further expansion of the United States offensive forces.

It had to be stressed, however, that the United States government did not know whether the Soviet Union would develop and deploy the kind of forces assumed, and even against this higher-than-expected threat, and even without a Nike-X defence of the Minuteman missiles, the proposed strategic missile and bomber forces could still inflict 40 per cent or more fatalities on the Soviet population throughout the time-period involved.

If the Soviet Union wished to reduce the vulnerability of their own offensive forces against the possibility of a first strike by America's own 'very accurate' forces in fiscal years 1972–3, they must further disperse and harden their strategic missiles, as they appeared to be doing at present: this was expensive, and for a given budget resulted in reduced missile payloads, but to neglect it would leave the Soviet force highly vulnerable.

The three main purposes of an ABM defence system such as Nike-X were (*a*) to protect the population and industry of United States cities against a Soviet missile attack; (*b*) to protect United States cities against a Chinese missile attack in the mid-

1970s; and (*c*) to help protect the United States land-based strategic offensive forces (i.e. Minuteman) against a Soviet missile attack. It was the conclusion of the government after exhaustive study of the subject and consideration of the views of its principal military and civilian advisers, that the United States 'should not initiate an ABM deployment at this time for any of these purposes'.

In particular, the Soviet Union would be forced to react to a United States ABM deployment by increasing its offensive nuclear force still further, so that the risk of a Soviet nuclear attack on the United States would not be further decreased and the damage from a Soviet nuclear attack, if deterrence were to fail, would 'not be reduced in any meaningful sense'. Deterrence was the foundation of United States security, and a Soviet nuclear attack could be prevented if the Soviet Union understood that the United States possessed strategic nuclear forces 'so powerful as to be capable of absorbing a Soviet first strike and surviving with sufficient strength to impose unacceptable damage on them'. Such a power was possessed at present and must be maintained in the future.

From the beginning of the Nike-Zeus programme in 1955 to mid-1967, $4,000,000,000 would have been invested in ballistic missile defence research, including Mike-Zeus, Nike-X, and project Defender, while during the past five or six years $1,200,000,000 had been spent on the development of penetration aids. However, as the Soviet Union had 'essentially the same requirement for a deterrent or "assured destruction" force' as the United States, the deployment by the United States of an ABM defence which would 'degrade the destruction capability of the Soviet's offensive force to an unacceptable level' would lead to the expansion of that force, and 'this would leave us no better off than we were before'.

*United States Programme*

The proposed programme for the United States is:

(*a*) To pursue with undiminished vigour the development, test and evaluation of the Nike-X system, for which purpose a total of about $440,000,000 had been included in the Budget for fiscal year 1968, but to take no action at present to deploy the system.

(*b*) To initiate negotiations with the Soviet Union designed, through formal or informal agreement, to limit the deployment of anti-ballistic missile systems.

(*c*) To reconsider the deployment decision in the event of these discussions proving unsuccessful. Approximately $375,000,000 had been included in the Budget for fiscal year 1968 to provide for such actions as might be required at that time – e.g. the production of Nike-X for the defence of the United States offensive weapons system.

*Strategic Missile Forces*

The United States response to a Soviet deployment of an ABM defence would be the incorporation of appropriate penetration aids in the United States strategic missiles, with an assured destruction capability being maintained, against area defence interceptors, at a cost of less than 10 per cent of the cost of an ABM defence to the Soviet Union.

Against a combined Soviet expanded strategic missile and ABM threat, the most efficient alternative available would be to develop Poseidon with the new penetration aids and to 'retrofit' it into Polaris vessels.

*Minuteman*

In 1966 it had been planned to have an ultimate force of 1,000 mixed Minuteman-2s and 3s, but it had now been decided to raise the proportion of Minuteman-2s and to increase their payload, so that they could carry more penetration aids to counter an ABM defence. At the time the Minuteman-2s would be re-equipped with improved re-entry vehicles and 'penetration aid packages', while eventually the earliest Minuteman-2s would be replaced because of their age and a new improved re-entry vehicle would be developed for the Minuteman-3.

*Polaris/Poseidon*

Thirty-nine of the forty-one Polaris submarines planned would have become operational by mid-1967 and the final two submarines would be deployed by September 1967. At 30 June 1967, thirty-two of the submarines would actually be deployed, with six undergoing overhaul and one being converted from A-1 to A-3 Polaris missiles; when this last named had completed its refit and the final two submarines had been deployed, there would be thirteen ships equipped with A-2s and 28 with A-3s.

It was now planned to produce and deploy Poseidon missiles. Although the cost of converting a submarine to Poseidon, of procuring new missiles, and of ten years of operation was about half as much again as that of operating a Polaris submarine for ten years, the effectiveness of the Poseidon submarine was several times greater. Of the total extra cost of developing Poseidon and of producing and deploying the proposed force ($3,300,000,000), about $900,000,000 was included in the Budget for fiscal year 1968, although this decision would produce an offsetting saving of about $200,000,000 in the Polaris programme.

*Titan-2*

The Titan-2 force comprised fifty-four missiles in silos. Although its range of 6,100 nautical miles was greater than that of the existing Minuteman and Polaris missiles, with the development of the Minuteman-3 and Poseidon it would 'no longer be unique'. The Titan-2 was, moreover, expensive to operate (probably nearly $1,000,000 per missile annually, including indirect costs), and the procurement of new Titan boosters for testing and operational reliability demonstration would be ended. With about six follow-on tests each year, the force of fifty-four Titan missiles on launchers could be maintained for a number of years.

*New Strategic Missiles*

A 'very comprehensive study' was being made of a new long-range missile system; in order to shorten the lead-time on any option selected as a result of this study, funds had been included in the Budget for fiscal year 1968 for contract definition should such a decision become warranted.

*Strategic Bomber Forces*

The manned bomber forces which would be maintained up to 1972 would be 255 B-52G-Hs and 210 FB-111As, while the B-52C-Fs and B-58s would be phased out.

### (*b*) MCNAMARA'S SAN FRANCISCO SPEECH

In a speech at San Francisco in September 1967, Mr McNamara announced the American Government's decision to deploy a 'thin' ABM system designed primarily to protect the United States against a hypothetical Chinese thermonuclear attack in the mid-1970s. The following quotations are from this speech:

*Gross Megatonnage Misleading*

Many commentators on the matter tend to define nuclear superiority in terms of gross megatonnage, or in terms of the number of missile launchers available. Now, by both these two standards of measurement, the United States does have a substantial superiority over the Soviet Union in the weapons targeted against each other.

But it is precisely these two standards of measurement that are themselves misleading. For the most meaningful and realistic measurement of nuclear capability is neither gross megatonnage nor the number of available missile launchers, but rather the number of separate warheads that are capable of being delivered with accuracy on individual high-priority targets with sufficient power to destroy them.

Gross megatonnage in itself is an inadequate indicator of assured destruction capability, since it is unrelated to survivability, accuracy, or penetrability, and poorly related to effective elimination of multiple high-priority targets. There is manifestly no advantage in over-destroying one target, at the expense of leaving undamaged other targets of equal importance.

Further, the number of missile launchers available is also an inadequate indicator of assured destruction capability, since the fact is that many of our launchers will carry multiple warheads.

I want, however, to make one point patently clear; our current numerical superiority over the Soviet Union in reliable, accurate, and effective warheads is both greater than we had originally planned, and is in fact more than we require. Moreover, in the large equation of security, our 'superiority' is of limited significance – since even with our current superiority, or indeed with any numerical superiority realistically attainable, the blunt, inescapable fact remains that the Soviet Union could still – with it present forces – effectively destroy the United States, even after absorbing the full weight of an American first strike.

*Arms Race Implications*

What is essential to understand is that the Soviet Union and the United States mutually influence one another's strategic plans. Whatever be their intentions, whatever be our intentions, actions – or even realistically potential actions, on either side relating to the build-up of nuclear forces, be they either offensive or defensive weapons, necessarily trigger reactions on the other side. It is precisely this action-reaction phenomenon that fuels an arms race.

Now, in strategic nuclear weaponry, the arms race involves a particular irony. Unlike any other era in military history, today a substantial numerical superiority of weapons does not effectively translate into political control or diplomatic leverage.

While thermonuclear power is almost inconceivably awesome and represents virtually unlimited potential destructiveness, it has proven to be a limited diplomatic instrument. Its uniqueness lies in the fact that it is, at one and the same time, an all-powerful weapon and a very inadequate weapon.

We do not want a nuclear arms race with the Soviet Union – primarily because the action-reaction phenomenon makes it foolish and futile. But if the only way to prevent the Soviet Union from obtaining first-strike capability over us is to engage in such a race, the United States possesses in ample abundance the resources, the technology, and the will to run faster in that race for whatever distance is required.

But what we would much prefer to do is to come to a realistic and reasonably riskless agreement with the Soviet Union which would effectively prevent such an arms race. We both have strategic nuclear arsenals greatly in excess of a credible assured destruction capability. These arsenals have reached that point of excess in each case for precisely the same reason – we have each reacted in the other's build-up with very conservative calculations. We have, that is, each built a greater arsenal than either of us needed for a second-strike capability, simply because we each wanted to be able to cope with the 'worst plausible case'.

But since we now each possess a deterrent in excess of our individual needs, both of our nations would benefit from a properly safeguarded agreement first to limit, and later to reduce, both our offensive and defensive strategic nuclear forces. We may, or we may not, be able to achieve such an agreement. We hope we can. And we believe such an agreement is fully feasible, since it is clearly in both our nations' interests.

But reach formal agreement or not, we can be sure that neither the Soviets nor we are going to risk the other's obtaining a first-strike capability. On the contrary, we can be sure that we are both going to maintain a maximum effort to preserve an assured destruction capability.

It would not be sensible for either side to launch a maximum effort to achieve a first-strike capability. It would not be sensible because the intelligence-gathering capability of each side being what it is, and the realities of lead-time from technological breakthrough to operational readiness being what they are, neither of us would be able to acquire a first-strike capability in secret.

*Reaction to Soviet ABM system*

The Soviets are now deploying an anti-ballistic missile system. If we react to this deployment intelligently, we have no reason for alarm. The system does not impose any threat to our ability to penetrate and inflict massive and unacceptable damage on the Soviet Union. In other words, it does not presently affect in any significant manner our assured destruction capability.

It does not impose such a threat because we have already taken the steps necessary to assure that our land-based Minuteman missiles, our nuclear submarine-launched new Poseidon missiles, and our strategic bomber forces have the requisite penetration aids – and, in sum, constitute a force of such magnitude that they guarantee us a force strong enough to survive a Soviet attack and penetrate the Soviet ABM deployment.

Now let me come to the issue that has received so much attention recently – the question of whether or not we should deploy an ABM system against the Soviet nuclear threat.

To begin with, this is not in any sense a new issue. We have had both the technical possibility and the strategic desirability of an American ABM deployment under constant review since the late 1950s.

While we have substantially improved our technology in this field, it is important to understand that none of the systems at the present or foreseeable state of the art would provide an impenetrable shield over the United States. Were such a shield possible, we would certainly want it – and we would certainly build it——

It has been alleged that we are opposed to deploying a large-scale ABM system because it would carry the heavy price-tag of $40,000,000,000. Let me make very clear that the $40,000,000,000 is not the issue. If we could build and deploy a genuinely impenetrable shield over the United States, we would be willing to spend not $40,000,000,000, but any reasonable multiple of that amount that was necessary.

The money in itself is not the problem – the penetrability of the proposed shield is the problem. There is clearly no point, however, in spending $40,000,000,000 if it is not going to buy us a significant improvement in our security——

Every ABM system that is now feasible involves firing defensive missiles at incoming offensive warheads in an effort to destroy them. But what many commentators overlook is that any such system can rather obviously be defeated by an enemy simply sending more offensive warheads, or dummy warheads, than there are defensive missiles capable of disposing of them.

This is the whole crux of the nuclear action-reaction phenomenon. Were we to deploy a heavy ABM system throughout the United States, the Soviets would clearly be strongly motivated to so increase their offensive capability as to cancel out any defensive advantage.

It is futile for each of us to spend $4,000,000,000, $40,000,000,000 or $400,000,000,000 – and at the end of all the spending, and all the deployment, and all the effort, to be relatively at the same point of balance on the security scale that we are now.

In point of fact, we have already initiated offensive weapons programmes costing several billions (thousand millions) in order to offset the small present Soviet ABM deployment, and the possibly more extensive future Soviet ABM deployments. That is money well spent: and it is necessary——

Keeping in mind the careful clockwork of lead-time, we will be forced to continue that effort over the next few years if the evidence is that the Soviets intend to turn what is now a light and modest ABM deployment into a massive one. Should they elect to do so, we have both the lead-time and the technology available to so increase both the quality and quantity of our offensive strategic forces – with particular attention to highly reliable penetration aids – that their expensive defensive efforts will give them no edge in the nuclear balance whatever.

But we would prefer not to have to do that. For it is a profitless waste of resources, provided we and the Soviets can come to a realistic strategic arms-limitation agreement. As you know, we have proposed American–Soviet talks on this matter. Should these talks fail, we are fully prepared to take the appropriate measures that such a failure would make necessary.

The point for us to keep in mind is that should the talks fail, and the Soviets decide to expand their present modest ABM deployment into a massive one, our response must be realistic. There is no point whatever in our responding by going to a massive ABM deployment to protect our population, when such a system would be ineffective against a sophisticated Soviet offence. Instead, realism dictates that if the Soviets elect to deploy a heavy ABM system, we must further expand our sophisticated offensive forces, and thus preserve our overwhelming assured destruction capability.

But the intractable fact is that should the talks fail, both the Soviets and ourselves would be forced to continue on a foolish and feckless course. It would be foolish and feckless because, in the end, it would provide neither the Soviets nor us with any greater relative nuclear capability. The time has come for us both to realize that, and to act reasonably. It is clearly in our own mutal interest to do so.

Having said that, it is important to distinguish between an ABM system designed to protect against a Soviet attack on our cities, and ABM systems which have other objectives.

*Chinese-Oriented ABM Systems*

One of the other uses of an ABM system which we should seriously consider is the greater protection of our strategic offensive forces. Another is in relation to the emerging nuclear capability of Communist China.

There is evidence that the Chinese are devoting very substantial resources to the development of both nuclear warheads and missile delivery systems. As I stated last January, indications are that they will have medium-range ballistic missiles within a year or so, an initial intercontinental ballistic missile capability in the early 1970s and a modest force in the mid-70s.

Up to now, the lead-time factor has allowed us to postpone a decision on whether or not a light ABM deployment might be advantageous as a counter-measure to Communist China's nuclear development. But the time will shortly be right for us to initiate production if we desire such a system.

President Johnson has made it clear that the United States will oppose any efforts of China to employ nuclear blackmail against her neighbours. We possess now, and will continue to possess for as far ahead as we can foresee, an overwhelming first-strike capability with respect to China. And despite the shrill and raucous propaganda directed at her own people that 'the atomic bomb is a paper tiger', there is ample evidence that

China well appreciates the destructive power of nuclear weapons.

China has been cautious to avoid any action that might end in a nuclear clash with the United States – however wild her words – and understandably so. We have the power not only to destroy completely her entire nuclear offensive forces, but to devastate her society as well.

Is there any possibility, then, that by the mid-1970s China might become so incautious as to attempt a nuclear attack on the United States or our allies? It would be insane and suicidal for her to do so, but one can conceive conditions under which China might miscalculate. We wish to reduce such possibilities to a minimum.

And since, as I have noted, our strategic planning must always be conservative, and take into consideration even the possible irrational behaviour of potential adversaries, there are marginal grounds for concluding that a light deployment of U.S. ABMs against this possibility is prudent.

The system would be relatively inexpensive – preliminary estimates place the cost at about $5,000,000,000 – and would have a much higher degree of reliability against a Chinese attack than the much more massive and complicated system that some have recommended against a possible Soviet attack.

Moreover, such an ABM deployment designed against a possible Chinese attack would have a number of other advantages. It would provide an additional indication to Asians that we intend to deter China from nuclear blackmail, and thus would contribute toward our goal of discouraging nuclear weapon proliferation among the present non-nuclear countries.

Further, the Chinese-oriented ABM deployment would enable us to add as a concurrent benefit a further defense of our Minuteman sites against Soviet attack, which means that at modest cost we would in fact be adding even greater effectiveness to our offensive missile force and avoiding a much more costly expansion of that force.

Finally, such a reasonably reliable ABM system would add protection of our population against the improbable but possible accidental launch of an intercontinental missile by any of the nuclear powers.

After a detailed review of all these considerations, we have decided to go forward with this Chinese-oriented ABM deployment, and we will begin actual production of such a system at the end of this year.

*Dangers of United States ABM Deployment*

I want to emphasize that it contains two possible dangers – and we should guard carefully against each.

The first danger is that we may psychologically lapse into the old over-simplification about the adequacy of nuclear power. The simple truth is that nuclear weapons can serve to deter only a narrow range of threats. This ABM deployment will strengthen our defensive posture, and will enhance the effectiveness of our land-based ICBM offensive forces. But the independent nations of Asia must realize that these benefits are no substitute for their maintaining and, where necessary, strengthening their own conventional forces in order to deal with more likely threats to the security of the region.

The second danger is also psychological. There is a kind of mad momentum intrinsic to the development of all new nuclear weapons. If a weapon system works – and works well – there is strong pressure from many directions to produce and deploy the weapon out of all proportion to the prudent level required.

The danger in deploying this relatively light and reliable Chinese-oriented ABM system is going to be that pressures will develop to expand it into a heavy Soviet-oriented ABM system. We must resist that temptation firmly – not because we can for a moment afford to relax our vigilance against a possible Soviet first strike – but precisely because our greatest deterrent against such a strike is not a massive, costly, but highly penetrable ABM shield, but rather a fully credible offensive assured destruction capability.

The so-called heavy ABM shield – at the present state of technology – would in effect be no adequate shield at all against a Soviet attack, but rather a strong inducement for the Soviets to vastly increase their own offensive forces. That, as I have pointed out, would make it necessary for us to respond in turn – and so the arms race would rush hopelessly on to no sensible purpose on either side.

Let me emphasize – and I cannot do so too strongly – that our decision to go ahead with limited ABM deployment in no way indicates that we feel an agreement with the Soviet Union on the limitation of strategic nuclear offensive and defensive forces is any the less urgent or desirable.

## ABBREVIATIONS USED IN TEXT

| | |
|---|---|
| ABM | Anti-Ballistic Missile. |
| BAMBI | Ballistic Missile Booster Interceptor. |
| BMEWS | Ballistic Missile Early Warning System. |
| BUIC | Back-up Interceptor Control. |
| CD | Civil Defence. |
| CRM | Cruise Missile. |
| CTBT | Comprehensive Test Ban Treaty. |
| DEW | Distant Early Warning Line. |
| DOD | United States Department of Defense. |
| FOBS | Fractional Orbital Ballistic System. |
| GCD | General and Complete Disarmament. |
| HIBEX | High-acceleration Booster Experiment. |
| ICBM | Inter-continental Ballistic Missile. |
| IRBM | Intermediate Range Ballistic Missile. |
| LGM | Silo-launched Missile. |
| MAR | Multi-function Array Radar. |
| MGM | Mobile-guided Missile. |
| MIDAS | Missile Detection and Alarm System. |
| MIRV | Multiple Independently Targetable Re-entry Vehicle. |
| MRBM | Medium Range Ballistic Missile. |
| MSR | Missile Site Radar. |
| NPT | Non-proliferation Treaty. |
| NREC | National Resource Evaluation Center (U.S.). |
| OCD | Office of Civil Defense (U.S.). |
| OCDM | Office of Civil Defense Mobilization. |
| ORNL | Oakridge National Laboratory. |
| PAR | Perimeter Acquisition Radar. |
| PARM | Program Analysis for Resource Management. |
| PTBT | Partial Test Ban Treaty. |
| SABMIS | Seaborne Anti-Ballistic Missile Intercept System. |
| SAGE | Semi Automatic Ground Environment System. |
| SLBM | Submarine-launched Cruise Missile. |
| SLM | Ship-launched Missile. |
| SRM | Short Range Missile |
| TACMAR | Tactical Multi-function Array Radar. |
| UGM | Underwater-launched Missile. |
| USAEC | United States Atomic Energy Commission. |